DSP for MATLAB™ and LabVIEW™

Volume I: Fundamentals of Discrete Signal Processing

SYNTHESIS LECTURES ON SIGNAL PROCESSING

Editor
José Moura, Carnegie Mellon University

DSP for MATLAB™ and LabVIEW™
Volume I: Fundamentals of Discrete Signal Processing
Forester W. Isen
2008

The Theory of Linear Prediction
P. P. Vaidyanathan
2007

Nonlinear Source Separation
Luis B. Almeida
2006

Spectral Analysis of Signals
Yanwei Wang, Jian Li, and Petre Stoica
2006

DSP for MATLAB™ and LabVIEW™ Volume I: Fundamentals of Discrete Signal Processing
Forester W. Isen

ISBN: 978-3-031-01400-0   paperback
ISBN: 978-3-031-02528-0   ebook

DOI 10.1007/978-3-031-02528-0

A Publication in the Springer series
SYNTHESIS LECTURES ON SIGNAL PROCESSING

Lecture #4
Series Editor: José Moura, Carnegie Mellon University

Series ISSN
Synthesis Lectures on Signal Processing
Print 1932-1236   Electronic 1932-1694

DSP for MATLAB™ and LabVIEW™
Volume I: Fundamentals of Discrete Signal Processing

Forester W. Isen

SYNTHESIS LECTURES ON SIGNAL PROCESSING #4

ABSTRACT

This book is Volume I of the series *DSP for MATLAB*™ *and LabVIEW*™. The entire series consists of four volumes that collectively cover basic digital signal processing in a practical and accessible manner, but which nonetheless include all essential foundation mathematics. As the series title implies, the scripts (of which there are more than 200) described in the text and supplied in code form (available via the internet at http://www.morganclaypool.com/page/isen will run on both MATLAB and LabVIEW. Volume I consists of four chapters. The first chapter gives a brief overview of the field of digital signal processing. This is followed by a chapter detailing many useful signals and concepts, including convolution, recursion, difference equations, LTI systems, etc. The third chapter covers conversion from the continuous to discrete domain and back (i.e., analog-to-digital and digital-to-analog conversion), aliasing, the Nyquist rate, normalized frequency, conversion from one sample rate to another, waveform generation at various sample rates from stored wave data, and Mu-law compression. The fourth and final chapter of the present volume introduces the reader to many important principles of signal processing, including correlation, the correlation sequence, the Real DFT, correlation by convolution, matched filtering, simple FIR filters, and simple IIR filters. Chapter 4, in particular, provides an intuitive or "first principle" understanding of how digital filtering and frequency transforms work, preparing the reader for Volumes II and III, which provide, respectively, detailed coverage of discrete frequency transforms (including the Discrete Time Fourier Transform, the Discrete Fourier Transform, and the z-Transform) and digital filter design (FIR design using Windowing, Frequency Sampling, and Optimum Equiripple techniques, and Classical IIR design). Volume IV, the culmination of the series, is an introductory treatment of LMS Adaptive Filtering and applications. The text for all volumes contains many examples, and many useful computational scripts, augmented by demonstration scripts and LabVIEW Virtual Instruments (VIs) that can be run to illustrate various signal processing concepts graphically on the user's computer screen.

KEYWORDS

Higher-Level Terms:
MATLAB, LabVIEW, DSP (Digital Signal Processing), Sampling, LTI Systems, Analog-to-Digital, Digital-to-Analog, FIR, IIR, DFT, Time Domain, Frequency Domain, Aliasing, Binary Numbers.
Lower-Level Terms:
Correlation, Convolution, Matched Filtering, Orthogonality, Interpolation, Decimation, Mu-Law, Stability, Causality, Difference Equations, Zero-Order Hold.

This volume is dedicated to the following memorable teachers of mine

Louise Costa
Cdr. Charles Bradimore Brouillette
Rudd Crawford
Sheldon Sarnevitz
Dr. C. W. Rector
Dr. Samuel Saul Saslaw
Dr. R. D. Shelton

Contents

Preface to Volume I

0.1 INTRODUCTION

The present volume is Volume I of the series *DSP for MATLAB*™ *and LabVIEW*™. The entire series consists of four volumes which collectively form a work of twelve chapters that cover basic digital signal processing in a practical and accessible manner, but which nonetheless include the necessary foundation mathematics. The text is well-illustrated with examples involving practical computation using m-code or MathScript (as m-code is usually referred to in LabVIEW-based literature), and LabVIEW VIs. There is also an ample supply of exercises, which consist of a mixture of paper-and-pencil exercises for simple computations, and script-writing projects having various levels of difficulty, from simple, requiring perhaps ten minutes to accomplish, to challenging, requiring several hours to accomplish.

As the series title implies, the scripts given in the text and supplied in code form (available via the internet at **http://www.morganclaypool.com/page/isen**) are suitable for use with both MATLAB (a product of The Mathworks, Inc.), and LabVIEW (a product of National Instruments, Inc.). Appendix A in each volume of the series describes the naming convention for the software written for the book as well as basics for using the software with MATLAB and LabVIEW.

0.2 THE FOUR VOLUMES OF THE SERIES

Volume I consists of four chapters. The first chapter gives a brief overview of the field of digital signal processing. This is followed by a chapter detailing many useful signals and concepts, including convolution, recursion, difference equations, etc. The third chapter covers conversion from the continuous to discrete domain and back (i.e., analog-to-digital and digital-to-analog conversion), aliasing, the Nyquist rate, normalized frequency, conversion from one sample rate to another, waveform generation at various sample rates from stored wave data, and Mu-law compression. The fourth and final chapter of the present volume introduces the reader to many important principles of signal processing, including correlation, the correlation sequence, the Real DFT, correlation by convolution, matched filtering, simple FIR filters, and simple IIR filters.

Volume II of the series is devoted to discrete frequency transforms. It begins with an overview of a number of well-known continuous domain and discrete domain transforms, and covers the DTFT (Discrete Time Fourier Transform), the DFT (Discrete Fourier Transform), and the z-Transform in detail. Filter realizations (or topologies) are also covered, including Direct, Cascade, Parallel, and Lattice forms.

Volume III of the series covers FIR and IIR design, including general principles of FIR design, the effects of windowing and filter length, characteristics of four types of linear-phase FIR, Comb and MA filters, Windowed Ideal Lowpass filter design, Frequency Sampling design with optimized transition band coefficients, Equiripple FIR design, and Classical IIR design.

Volume IV of the series, LMS Adaptive Filtering, begins by explaining cost functions and performance surfaces, followed by the use of gradient search techniques using coefficient perturbation, finally reaching the elegant and computationally efficient Least Mean Square (LMS) coefficient update algorithm. The issues of stability, convergence speed, and narrow-bandwidth signals are covered in a practical manner, with many illustrative scripts. In the second chapter of the volume, use of LMS adaptive filtering in various filtering applications and topologies is explored, including Active Noise Cancellation (ANC), system or plant modeling, periodic component elimination, Adaptive Line Enhancement (ADE), interference cancellation, echo cancellation, and equalization/deconvolution.

0.3   ORIGIN AND EVOLUTION OF THE SERIES

The manuscript from which the present series of four books has been made began with an idea to provide a basic course for intellectual property specialists and engineers that would provide more explanation and illustration of the subject matter than that found in conventional academic books. The idea to provide an accessible basic course in digital signal processing began in the mid-to-late 1990's when I was introduced to MATLAB by Dan Hunter, whose graduate school days occurred after the advent of both MATLAB and LabVIEW (mine did not). About the time I was seriously exploring the use of MATLAB to update my own knowledge of signal processing, Dr. Jeffrey Gluck began giving an in-house course at the agency on the topics of convolutional coding, trellis coding, etc., thus inspiring me to do likewise in the basics of DSP, a topic more in-tune to the needs of the unit I was supervising at the time. Two short courses were taught at the agency in 1999 and 2000 by myself and several others, including Dr. Hal Zintel, David Knepper, and Dr. Pinchus Laufer. In these courses we stressed audio and speech topics in addition to basic signal processing concepts. Thanks to The Mathworks, Inc., we were able to teach the in-house course with MATLAB on individual computers, and thanks to Jim Dwyer at the agency, we were able to acquire several server-based concurrent-usage MATLAB licenses, permitting anyone at the agency to have access to MATLAB. Some time after this, I decided to develop a complete course in book form, the previous courses having consisted of an ad hoc pastiche of topics presented in summary form on slides, augmented with visual presentations generated by custom-written scripts for MATLAB. An early draft of the book was kindly reviewed by Motorola Patent Attorney Sylvia Y. Chen, which encouraged me to contact Tom Robbins at Prentice-Hall concerning possible publication. By 2005, Tom was involved in starting a publishing operation at National Instruments, Inc., and introduced me to LabVIEW with the idea of possibly crafting a book on DSP to be compatible with LabVIEW. After review of a manuscript draft by a panel of three in early 2006, it was suggested that all essential foundation mathematics be included so the book would have both academic and professional appeal. Fortunately, I had long since retired from the agency and was able to devote the considerable amount of time needed for such a project. The result is a book suitable for use in both academic and professional settings, as it includes essential mathematical formulas and concepts as well as simple or "first principle" explanations that help give the reader a gentler entry into the more conventional mathematical treatment.

This double-pronged approach to the subject matter has, of course, resulted in a book of considerable length. Accordingly, it has been broken into four modules or volumes (described above) that together form a comprehensive course, but which may be used individually by readers who are not in need of a complete course.

Many thanks go not only to all those mentioned above, but to Joel Claypool of Morgan & Claypool, Dr. C.L. Tondo and his troops, and, no doubt, many others behind the scenes whose names I have never heard, for making possible the publication of this series of books.

Forester W. Isen
November 2008

CHAPTER 1

An Overview of DSP

1.1 SIGNALS, WAVES, AND DIGITAL PROCESSING

Two of the human senses, sight and hearing, work via the detection of waves. Useful information from both light and sound is gained by detection of certain characteristics of these waves, such as frequency and amplitude. Modern telecommunication depends on transducing sound or light into electrical quantities such as voltage, and then processing the voltage in many different ways to enable the information to be reliably stored or conveyed to a distant place and then regenerated to imitate (i.e., reconstruct) the original sound or light phenomenon.

For example, in the case of sound, a microphone detects rapid pressure variations in air and converts those variations to an output voltage which varies in a manner proportional to the variation of pressure on the microphone's diaphragm. The varying voltage can be used to cut a corresponding wave into a wax disc, to record corresponding wave-like variations in magnetism onto a ferromagnetic wire or tape, to vary the opacity of a linear track along the edge of a celluloid film (i.e., the sound-track of a motion picture film) or perhaps to modulate a carrier wave for radio transmission.

In recent decades, signal processing and storage systems have been developed that use discrete samples of a signal rather than the entire continuous time domain (or **analog**) signal. Several useful definitions are as follows:

- A **sample** is the amplitude of an analog signal at an instant in time.

- A system that processes a signal in sampled form (i.e., a sequence of samples) is known as a **Discrete Time Signal Processing System**.

- In a **Digital Signal Processing** system, the samples are converted to numerical values, and the values (numbers) stored (usually in binary form), transmitted, or otherwise processed.

The difference between conventional analog systems and digital systems is illustrated in Fig. 1.1. At (a), a conventional analog system is shown, in which the signal from a microphone is sent directly to an analog recording device, such as a tape recorder, recorded at a certain tape speed, and then played back at the same speed some time later to reproduce the original sound. At (b), samples of the microphone signal are obtained by an **Analog-to-Digital Converter (ADC)**, which converts instantaneous voltages of the microphone signal to corresponding numerical values, which are stored in a digital memory, and can later be sent to a **Digital-to-Analog Converter (DAC)** to reconstruct the original sound.

In addition to recording and reproducing analog signals, most other kinds of processing which might be performed on an analog signal can also be performed on a sampled version of the signal

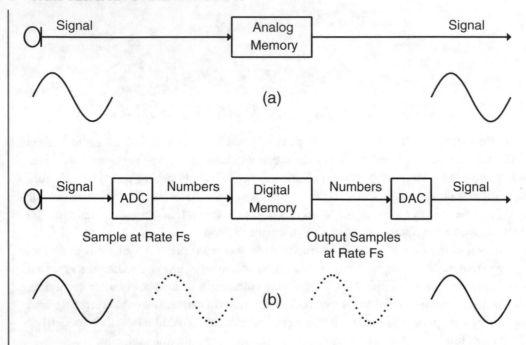

Figure 1.1: (a) Conventional analog recording and playback system; (b) A digital recording and playback system.

by using numerical algorithms. These can be categorized into two broad types of processing, time domain and frequency domain, which are discussed in more detail below.

1.2   ADVANTAGES OF DIGITAL PROCESSING

The reduction of continuous signals to sequences of numerical values (samples) that can be used to process and/or reconstruct the original signal, provides a number of benefits that cannot be achieved with continuous or analog signal processing. The following are some of the benefits of digital processing:

1. Analog hardware, such as amplifiers, filters, comparators, etc., is very susceptible to noise and deterioration through aging. Digital hardware works with only two signal levels rather than an infinite number, and hence has a high signal to noise ratio. As a result, there is little if any gradual deterioration of performance with age (although as with all things, digital hardware can suddenly and totally fail), and copies of signal files are generally perfect, absent component failure, media degeneration or damage, etc. This is not true with analog hardware and recording techniques, in which every copy introduces significant amounts of additional noise and distortion.

2. Analog hardware, for the most part, must be built for each processing function needed. With digital processing, each additional function needed can be implemented with a software module,

using the same piece of hardware, a digital computer. The computing power available to the average person has increased enormously in recent years, as evidenced by the incredible variety of inexpensive, high quality devices and techniques available. Hundreds of millions or billions of operations per second can be performed on a signal using digital hardware at reasonable expense; no reasonably-priced alternative exists using analog hardware and processing.

   3. Analog signal storage is typically redundant, since wave-related signals (audio, video, etc.) are themselves typically redundant. For example, by taking into account this redundancy as well as the physiological limitations of human hearing, storage needs for audio signals can be reduced up to about 95%, using digitally-based compression techniques, such as MP3, AC3, AAC, etc.

   4. Digital processing makes possible highly efficient security and error-correction coding. Using digital coding, it is possible, for example, for many signals to be transmitted at very low power and to share the same bandwidth. Modern cell phone techniques, such as CDMA (Code Division Multiple Access) rely heavily on advanced, digitally-based signal processing techniques to efficiently achieve both high quality and high security.

1.3 DSP NOMENCLATURE AND TOPICS

Figure 1.2 shows a broad overview of digital signal processing. Analog signals enter an ADC from the left, and samples exit the ADC from the right, and may be 1) processed strictly in the discrete time domain (in which samples represent the original signal at instants in time) or they may be 2) converted to a frequency domain representation (in which samples represent amplitudes of particular frequency components of the original signal) by a time-to-frequency transform, processed in the frequency domain, then converted back to the discrete time domain by a frequency-to-time transform. Discrete time domain samples are converted back to the continuous time domain by the DAC.

   Note that a particular signal processing system might use only time domain processing, only frequency domain processing, or both time and frequency domain processing, so either or both of the signal processing paths shown in Fig. 1.2 may be taken in any given system.

1.3.1 TIME DOMAIN PROCESSING

Filtering, in general, whether it is done in the continuous domain or discrete domain, is one of the fundamental signal processing techniques; it can be used to separate signals by selecting or rejecting certain frequencies, enhance signals (such as with audio equalization, etc.), alter the phase characteristic, and so forth. Hence a major portion of the study of digital signal processing is devoted to digital filtering. Filtering in the continuous domain is performed using combinations of components such as inductors, capacitors, resistors, and in some cases active elements such as op amps, transistors, etc. Filtering in the discrete or digital domain is performed by mathematically manipulating or processing a sequence of samples of the signal using a discrete time processing system, which typically consists of registers or memory elements, delay elements, multipliers, and adders. Each of the preceding elements may be implemented as distinct pieces of hardware in an efficient arrangement designed to function for a particular purpose (often referred to as a **Pipeline**

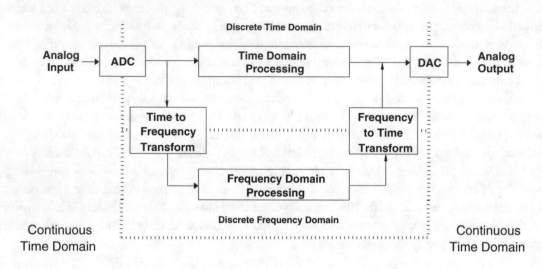

Figure 1.2: A broad, conceptual overview of digital signal processing.

Processor), or, the equivalent functions of all elements may be implemented on a general purpose computer by specifically designed software.

1.3.2   FREQUENCY TRANSFORMS

A time-to-frequency transform operates on a block of time domain samples and evaluates the frequency content thereof. A set of frequency coefficients is derived which can be used to quantify the amplitudes (and usually phases) of frequency components of the original signal, or the coefficients can be used to reconstruct the original time domain samples using an inverse transform (a frequency-to-time transform). The most well-known and widely-used of these transforms is the **Discrete Fourier Transform (DFT)**, usually implemented by the **FFT** (for **Fast Fourier Transform**), the name of a class of algorithms that allow efficient computation of the DFT.

1.3.3   FREQUENCY DOMAIN PROCESSING

Most signal processing that can be done in the time domain can be also equivalently done in the frequency domain. Each domain has certain advantages for a given type of problem.

Time domain filtering, for example, can be performed using frequency transforms such as the DFT, and in certain cases efficiency can be greatly improved using this technique.

A second use is in digital filter design, in which the desired filter frequency response is specified in the frequency domain, i.e., as a set of DFT coefficients, for example.

Yet a third and very prevalent use is **Transform Coding**, in which signals are coded using a frequency transform (usually eliminating as much redundant information as possible) and then reconstructed from the transform coefficients. Transform Coding is a powerful tool for compression algorithms, such as those employed with MP3 (MPEG II, Level 3) for audio signals, JPEG, a common image compression format, etc. The use of such compression algorithms has revolutionized the audio and video fields, making storage of audio and video data very economical and deliverable via Internet.

1.4   ORGANIZATION OF THIS VOLUME OF THE SERIES

The present volume provides basic information on digital signal processing and has four chapters as follows:

- Chapter 1 (the present chapter) gives a brief overview of DSP. It defines sampling, contrasts the areas of continuous and discrete signal processing, as well as time domain and frequency domain processing, and introduces very basic signal processing nomenclature.

- Chapter 2 introduces many useful signals and sequences, followed by a basic introduction to Linear, Time Invariant (LTI) systems, including convolution, stability and causality, basic FIR and IIR filters, and difference equations.

- Chapter 3 discusses the fundamental concepts of sampling, analog-to-digital conversion, and digital-to-analog conversion. The topics of aliasing, normalized frequency, binary formats, zero-order hold conversion, interpolation, decimation, frequency generation, and Mu-law compression are also covered.

- Chapter 4 introduces correlation and the correlation sequence, orthogonality of sinusoids and complex exponentials, sequence decomposition and reconstruction using correlation (i.e., the real DFT), correlation via convolution, matched filtering, simple FIR filters, and the basic IIR using a single pole or complex conjugate pole pairs.

1.5   CONTENTS OF VOLUME II IN THE SERIES

The second book in the series covers standard digital frequency transforms and closely related topics.

- Chapter 1 begins with a short overview of the Fourier and Laplace families of transforms, calling attention to the uses of each and the differences among them. The remainder of the chapter is devoted to the Discrete Time Fourier Transform (DTFT), its properties, and its use in evaluation of the frequency response of an LTI system.

- Chapter 2 introduces the z-transform, its properties, the inverse z-transform, transfer function and various filter topologies (Direct, Cascade, Parallel, and Lattice Forms), and evaluation of the frequency response of an LTI system using the z-transform.

- Chapter 3 covers the Discrete Fourier Transform (DFT), including the forward and reverse transforms, properties, the Fast Fourier Transform (FFT), the Goertzel Algorithm, periodic, cyclic, and linear convolution via the DFT, and DFT leakage. Computation of the IDFT (inverse DFT) via DFT, computation of the DFT via matrix, and computation of the DTFT via the DFT, are also discussed.

1.6 CONTENTS OF VOLUME III IN THE SERIES

The third book in the series is devoted to digital filter design.

- Chapter 1 gives an overview of FIR filtering principles in general, and linear phase filter characteristics in particular, and an overview of FIR design techniques.

- Chapter 2 covers FIR design via windowed ideal lowpass filter, frequency sampling with optimized transition coefficients (as implemented by inverse DFT as well as cosine/sine summation formulas), and equiripple design. Designs are performed for lowpass, highpass, bandpass, and bandstop filters, as well as Hilbert transformers and differentiators.

- Chapter 3 is devoted to classical IIR design, including design of digital IIR filters starting from analog prototype lowpass filters of the Butterworth, Chebyshev (I and II), and Elliptic types, transformations from lowpass to other passband types in the analog domain, and analog-to-digital filter transformation. The chapter concludes with a discussion of various filter design functions provided by MATLAB and LabVIEW.

1.7 CONTENTS OF VOLUME IV IN THE SERIES

The fourth book in the series provides an introduction to LMS adaptive filtering:

- Chapter 1 discusses cost functions, performance surfaces, coefficient perturbation to estimate the gradient in the method of steepest descent, and the LMS algorithm and its performance with signals of differing frequency content.

- Chapter 2 covers a number of standard uses for the LMS algorithm in adaptive FIR filtering systems, including active noise cancellation, echo cancellation, interference cancellation, periodic component enhancement or elimination, and deconvolution.

CHAPTER 2

Discrete Signals and Concepts

2.1 OVERVIEW

If the study of digital signal processing is likened to a story, this chapter can be viewed as an introduction of the main characters in the story–they are the various types of signals (or their sampled versions, called sequences) and fundamental processes that we will see time and time again. Acquiring a good working knowledge of these is essential to understanding the rest of the story, just as is knowing the characters in a novel.

In the first part of the chapter, we introduce discrete sequence notation and many standard test signals including the unit impulse, the unit step, the exponential sequence (both real and complex), the chirp, etc., and we learn to add and multiply sequences that are offset in time. Any serious study of digital signal processing relies heavily on the representation of sinusoids by the complex exponential, and hence this is covered in detail in the chapter. In the latter part of the chapter, we introduce the concepts of linear, time-invariant (LTI) systems, convolution, stability and causality, the FIR, the IIR, and difference equations.

By the end of this chapter, the reader will be prepared for the next chapter in the story of DSP, namely, the process and requirements for obtaining sequences via sampling, formatting sample values in binary notation, converting sequences back into continuous domain signals, and changing the sample rate of a sequence.

2.2 SOFTWARE FOR USE WITH THIS BOOK

The software files needed for use with this book (consisting of m-code (.m) files, VI files (.vi), and related support files) are available for download from the following website:

http://www.morganclaypool.com/page/isen

The entire software package should be stored in a single folder on the user's computer, and the full file name of the folder must be placed on the MATLAB or LabVIEW search path in accordance with the instructions provided by the respective software vendor (in case you have encountered this notice before, which is repeated for convenience in each chapter of the book, the software download only needs to be done once, as files for the entire series of four volumes are all contained in the one downloadable folder).

See Appendix A for more information.

2.3   DISCRETE SEQUENCE NOTATION

Digital Signal Processing must necessarily begin with a signal, and most signals, such as sound, images, etc., originate as continuous-valued (or analog) signals, and must be converted into a sequence of samples to be processed using digital techniques.

Figure 2.1 depicts a continuous-domain sine wave, with eight samples marked, sequentially obtained every 0.125 second. The signal values input to the ADC at sample times 0, 0.125, 0.25, 0.375, 0.5, 0.625, 0.75, etc., are 0, 0.707, 1, 0.707, 0, -0.707, -1, etc.

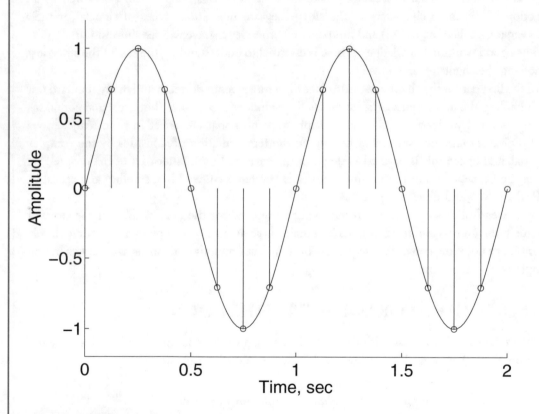

Figure 2.1: An analog or continuous-domain sine wave, with eight samples per second marked.

The samples within a given sample sequence are normally indexed by the numbers 0, 1, 2, etc., which represent multiples of the sample period T. For example, in Fig. 2.1, we note that the sample period is 0.125 second, and the actual sampling times are therefore 0 sec., 0.125 sec., 0.25 sec., etc. The continuous sine function shown has the value

$$f(t) = \sin(2\pi f t)$$

where t is time, f is frequency, and in this particular case, $f = 1$ Hz. Sampling occurs at times nT where n = 0, 1, 2,...and $T = 0.125$ second. The sample values of the sequence would therefore be $\sin(0)$, $\sin(2\pi(T))$, $\sin(2\pi(2T))$, $\sin(2\pi(3T))$, etc., and we would then say that $s[0] = 0$, $s[1] = 0.707$, $s[2] = 1.0$, $s[3] = 0.707$, etc. where $s[n]$ denotes the n-th sequence value, the amplitude of which is equal to the underlying continuous function at time nT (note that brackets surrounding a function argument mean that the argument can only assume discrete values, while parentheses surrounding an argument indicate that the argument's domain is continuous). We can also say that

$$s[n] = \sin[2\pi nT]$$

This sequence of values, the samples of the sine wave, can be used to completely reconstruct the original continuous domain sine wave using a DAC. There are, of course, a number of conditions to ensure that this is possible, and they are taken up in detail in the next chapter.

To compute and plot the sample values of a 2-Hz sine wave sampled every 0.05 second on the time interval 0 to 1.1 second, make the following MathScript call:

t = [0:0.05:1.1]; figure; stem(t,sin(2*pi*2*t))

where the t vector contains every sample time nT with $T = 0.05$. Alternatively, we might write

T = 0.05; n = 0:1:22; figure; stem(n*T,sin(2*pi*2*n*T))

both of which result in Fig. 2.2.

2.4   USEFUL SIGNALS, SEQUENCES, AND CONCEPTS

2.4.1   SINE AND COSINE

We saw above that a sine wave of frequency f periodically sampled at the time period T has the values

$$s[n] = \sin[2\pi fnT]$$

Once we have a sampled sine wave, we can mathematically express it without reference to the sampling period by defining the sequence length as N. We would then have, in general,

$$s[n] = \sin[2\pi nk/N]$$

where n is the sample index, which runs from 0 to $N-1$, and k is the number of cycles over the sequence length N. For the sample sequence marked in Fig. 2.1, we would have

$$s[n] = \sin[2\pi n2/16]$$

where we have noted that there are two full cycles of the sine over 16 samples (the 17th sample is the start of the third cycle). The correctness of this formula can be verified by noting that for

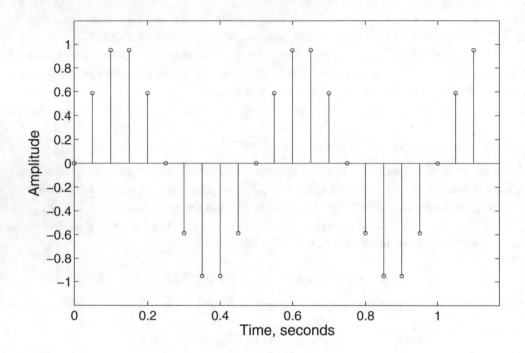

Figure 2.2: A plot of the samples of a sine wave having frequency 2 Hz, sampled every 0.05 second up to time 1.1 second.

the 17th sample, $n = 16$, and $s[16] = 0$, as shown. Picking another sample, for $n = 2$, we get $s[2] = \sin[2\pi(2)2/16] = \sin[\pi/2] = 1$, as shown.

A phase angle is sometimes needed, and is denoted θ by in the following expression:

$$s[n] = \sin[2\pi nk/N + \theta]$$

Note that if $\theta = \pi/2$, then

$$s[n] = \cos[2\pi nk/N]$$

We can illustrate this by generating and displaying a sine wave having three cycles over 18 samples, then the same sine wave, but with a phase angle of $\pi/2$ radians, and finally a cosine wave having three cycles over 18 samples and a zero phase angle. A suitable MathScript call, which results in Fig. 2.3, is

```
n = 0:1:17; y1 = sin(2*pi*n/18*3); subplot(311); stem(n,y1);
y2 = sin(2*pi*n/18*3 +pi/2); subplot(312); stem(n,y2);
y3 = cos(2*pi*n/18*3); subplot(313); stem(n,y3)
```

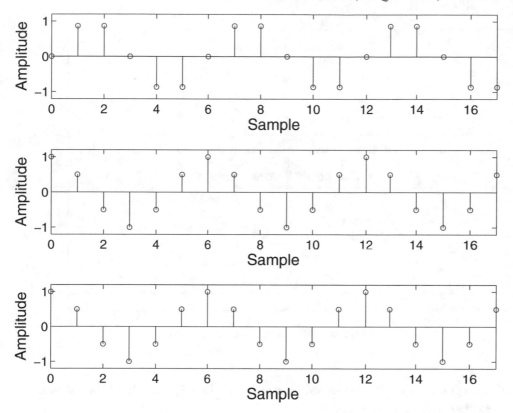

Figure 2.3: (a) Three cycles of a sine wave over 18 samples, with phase angle 0 radians; (b) Same as (a), with a phase angle of $\pi/2$ radians; (c) Three cycles of a cosine wave over 18 samples, with a phase angle of 0 radians.

2.4.2   SEQUENCE AND TIME POSITION VECTOR

Certain operations on two sequences, such as addition and multiplication, require that the sequences be of equal length, and that their proper positions in time be preserved.

Consider the sequence $x1 = [1,2,3,4]$, which was sampled at sample time indices $n1 = [-1,0,1,2]$, which we would like to add to sequence $x2 = [4,3,2,1]$, which was sampled at time indices $n2 = [2,3,4,5]$. To make these two sequences equal in length, we'll prepend and postpend zeros as needed to result in two sequences of equal length that retain the proper time alignment. We see that the minimum time index is -1 and the maximum time index is 5. Since $x1$ starts at the minimum time index, we postpend zeros to it such that we would have $x1 = [1,2,3,4,0,0,0]$, with corresponding time indices $[-1,0,1,2,3,4,5]$. Similarly, we prepend zeros so that $x2 = [0,0,0,4,3,2,1]$, with the same total time or sample index range as the modified version of $x1$. Figure 2.4 depicts this process.

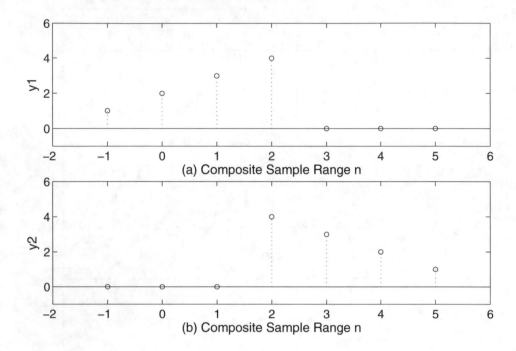

Figure 2.4: (a) First sequence, with postpended zeros at sample times 3, 4, and 5; (b) Second sequence, with prepended zeros at sample times -1, 0, and 1.

The sum is then

$$x1 + x2 = [1,2,3,4,0,0,0] + [0,0,0,4,3,2,1] = [1,2,3,8,3,2,1]$$

and has time indices [-1,0,1,2,3,4,5].

These two ideas, that sequences to be added or multiplied must be of equal length, but also properly time-aligned, lead us to write several MathScript functions that will automatically perform the needed adjustments and perform the arithmetic operation.

The following script will perform addition of offset sequences $y1$ and $y2$ that have respective time indices $n1$ and $n2$ using the method of prepending and postpending zeros.

```
function [y, nOut] = LVAddSeqs(y1,n1,y2,n2)
nOut = [min(min(n1),min(n2)):1:max(max(n1),max(n2))];
mnfv = min(nOut); mxfv = max(nOut);
y = [zeros(1,min(n1)-mnfv),y1,zeros(1,mxfv-max(n1))] + ...
[zeros(1,min(n2)-mnfv),y2,zeros(1,mxfv-max(n2))];
```

The function

$$[y, nOut] = LVMultSeqs(y1, n1, y2, n2)$$

works the same way, with the addition operator (+) in the final statement being replaced with the operator for multiplying two vectors on a sample-by-sample basis, a period following by an asterisk (.*).

We can illustrate use of the function $LVAddSeqs$ by using it to add the following sequences:

$$y1 = [3,-2,2], n1 = [-1,0,1], y2 = [1,0,-1], n2 = [0,1,2]$$

We make the call

$$[y, n] = LVAddSeqs([3,-2,2], [-1,0,1], [1,0,-1], [0,1,2])$$

which yields $y = [3,-1,2,-1]$ and $n = [-1,0,1,2]$.

We can illustrate use of the function $LVMultSeqs$ by using it to multiply the same sequences. We thus make the call

$$[y, n] = LVMultSeqs([3,-2,2], [-1,0,1], [1,0,-1], [0,1,2])$$

which yields $y = [0,-2,0,0]$ and $n = [-1,0,1,2]$.

2.4.3   THE UNIT IMPULSE (DELTA) FUNCTION

The **Unit Impulse** or **Delta Function is** defined as $\delta[n] = 1$ when $n = 0$ and 0 for all other values of n. The time of occurrence of the impulse can be shifted by a certain number of samples k using the notation $\delta[n - k]$ since the value of the function will only be 1 when $n - k = 0$.

The following function will plot a unit impulse at sample index n on the sample interval $Nlow$ to $Nhigh$.

```
function LVPlotUnitImpSeq(n,Nlow,Nhigh)
xIndices = [Nlow:1:Nhigh];
xVals = zeros(1,length(xIndices));
xVals(find(xIndices-n==0))=1;
stem(xIndices,xVals)
```

An example MathScript call is

$$LVPlotUnitImpSeq(-2,-10,10)$$

A version of the script that returns the output sequence and its indices without plotting is

$$[xVals,xIndices] = LVUnitImpSeq(n,Nlow,Nhigh)$$

This version is useful for generating composite unit impulse sequences. For example, we can display, over the sample index interval -5 to 5, the output sequence

$$y[n] = 3\delta[n-2] - 2\delta[n+3]$$

by using the following m-code, which computes and displays the desired output sequence using the function *LVUnitImpSeq*, as shown in Fig. 2.5.

```
[y1,y1Ind] = LVUnitImpSeq(2,-5,5),
[y2,y2Ind] = LVUnitImpSeq(-3,-5,5),
y = 3*y1 - 2*y2, stem(y1Ind,y)
```

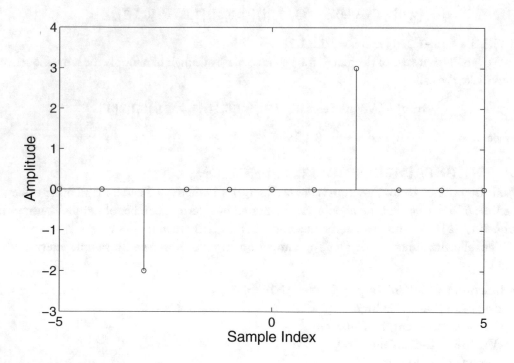

Figure 2.5:   A graph of the function $y[n] = 3[n-2] - 2[n+3]$ for sample indices -5 to +5.

2.4.4   THE UNIT STEP FUNCTION

The **Unit Step Function is** defined as $u[n] = 1$ when $n \geq 0$ and 0 for all other values of n. The time of occurrence of the step (the value 1) can be shifted by a certain number of samples k using the notation $u[n-k]$ since the value of the function will only be 1 when $n - k \geq 0$.

The following function will plot a unit step at sample index n on the sample interval *Nlow* to *Nhigh*.

```
function LVPlotUnitStepSeq(n,Nlow,Nhigh)
xIndices = [Nlow:1:Nhigh];
yVals(1:1:length(xIndices)) = 0;
posZInd = find((xIndices-n)==0)
yVals(posZInd:1:length(xIndices)) = 1;
stem(xIndices,yVals)
```

An example MathScript call is

$$\text{LVPlotUnitStepSeq}(-2,-10,10)$$

A version of the script that returns the output sequence and its indices without plotting is

$$[yVals, xIndices] = LVUnitStepSeq(n, Nlow, Nhigh)$$

This version is useful for generating composite unit step sequences. For example, we can display, over the sample index interval [-10:10] the sequence $y[n]$, defined as follows,

$$y[n] = 3u[n - 2] - 2u[n + 3]$$

with the following m-code, which computes and plots $y[n]$, using the function *LVUnitStepSeq*:

```
[y1,y1Ind] = LVUnitStepSeq(2,-10,10);
[y2,y2Ind] = LVUnitStepSeq(-3,-10,10);
y = 3*y1 - 2*y2; stem(y1Ind,y)
```

As another example, we can express the four-sample sequence [1,1,1,1] having time vector [-1,0,1,2] as a sum of unit step sequences and verify the answer using MathScript. To start, we generate a unit step sequence starting at $n = -1$ and subtract from it a unit step sequence starting at $n = 3$:

$$y = u[n + 1] - u[n - 3]$$

To verify, we can modify the code from the previous example; the results are shown in Fig. 2.6.

```
[y1,y1Ind] = LVUnitStepSeq(-1,-10,10);
[y2,y2Ind] = LVUnitStepSeq(3,-10,10);
y = y1 - y2; stem(y1Ind,y)
```

2.4.5   REAL EXPONENTIAL SEQUENCE
A signal generated as

$$y[n] = a^n$$

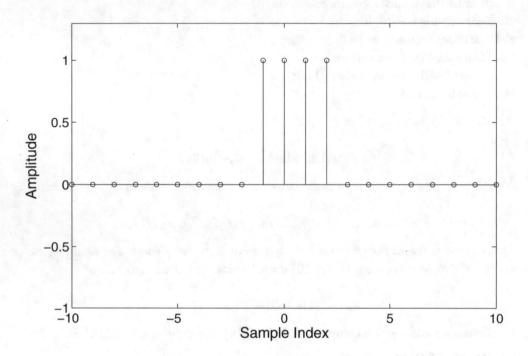

Figure 2.6: A plot over time indices -10 to +10 of the sequence defined as $y[n] = u[n + 1] - u[n - 3]$.

where a is a real number and n is real integer, produces a real sequence. In MathScript, to raise a number to a single power, use the "^" operator; to raise a number to a vector of powers, use the ".^" operator.

To illustrate this, we can generate and plot a real exponential sequence with a = 2 and n = [0:1:6]. A suitable MathScript call is

$$y = 2.\hat{}([0:1:6]); \text{figure; stem}(y)$$

As a second example, we generate and plot the real exponential sequence with a = 2 and n = [-6:1:0]. A suitable call is

$$n = [-6:1:0]; y = 2.\hat{}n; \text{figure; stem}(n,y)$$

the results of which are shown in Fig. 2.7.

2.4.6   PERIODIC SEQUENCES

A sequence that repeats itself exactly is called periodic. A periodic sequence can be generated from a given sequence S of length M by using the outer vector product of the sequence in column vector

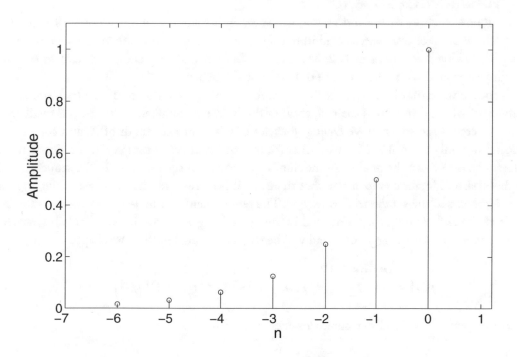

Figure 2.7: A real exponential sequence formed by raising the number 2 to the powers [-6:1:0].

form and a row vector of N ones. This generates an M-by-N matrix each column of which is the sequence S. The matrix can then be converted to a single column vector using MathScript's colon operator, and the resulting column vector is converted to a row vector by the transposition operator, the apostrophe.

The following function will generate n periods of the sequence y:

function nY = LVMakePeriodicSeq(y,N)
% LVMakePeriodicSeq([1 2 3 4],2)
y = y(:); nY = y*([ones(1,N)]); nY = nY(:)';

To illustrate use of the above, we will generate a sequence having three cycles of a cosine wave having a period of 11 samples. One period of the desired signal is

$$\cos(2*pi*[0:1:10]/11)$$

and a suitable call that computes and plots the result is therefore

N = 3; y = [cos(2*pi*[0:1:10]/11)]';
nY = LVMakePeriodicSeq(y,N); stem(nY)

2.4.7   HARMONIC SEQUENCES

Periodic signals, such as square and sawtoooth waves in a train, etc., are composed of sinusoids forming a harmonic (or overtone, as used in music) series. A periodic wave, then, is a superposition of sinusoids having frequencies of 1, 2, 3 ... times a fundamental frequency, with certain specific amplitudes associated with each overtone or harmonic frequency.

A square wave may be synthesized by summing sine waves having only odd harmonics, the amplitudes of which diminish as the reciprocal of the harmonic number. Equation (2.1) will synthesize a discrete time square wave having f cycles over a sequence length of N and containing odd harmonics only up to *MaxHar*. Note that N, the total number of samples in the signal, must be at least equal to twice the product of the number of cycles (of square wave) in the sequence and the highest harmonic number to ensure that there are at least two samples per cycle of the highest frequency present in the synthesized wave $y[n]$. The requirement that there be at least two samples per cycle of the highest frequency contained in the signal is a general one imposed on all signals by the properties of the sampling process, and will be studied in detail in the next chapter.

$$y[n] = \sum_{k=1}^{(MaxHar+1)/2} (1/(2k-1)) \sin(2\pi f(2k-1)(n/N)) \tag{2.1}$$

For a sawtooth wave, all harmonics are included:

$$y[n] = \sum_{k=1}^{MaxHar} (1/k) \sin(2\pi f k(n/N))$$

To illustrate the above formulas in terms of m-code, we will synthesize square and sawtooth waves having 10 cycles, up to the 99th harmonic.

We compute the necessary minimum sequence length as 2(10)(99) = 1980. The following MathScript call will synthesize a square wave up to the 99th harmonic; note that only odd harmonics are included, i.e., k = 1:2:99.

N=1980; n = 0:1:N; y = 0; for k = 1:2:99;
y = y + (1/k)*sin(2*pi*10*k*n/N); end; figure; plot(y)

For a sawtooth wave, the harmonic values are specified as 1:1:99, thus including both odd and even values:

N=1980; n = 0:1:N; y = 0; for k=1:1:99;
y = y + (1/k)*sin(2*pi*10*k*n/N); end; figure; plot(y)

The script

$$LV\,Synth\,Plot\,Square\,Sawtooth(WaveType, FundFreq, SR)$$

will synthesize square, sawtooth, and triangle waves one harmonic at a time (press any key for the next harmonic after making the initial call); specify *WaveType* as 1 for sawtooth, 2 for square, or 3 for triangle. The following call results in Fig. 2.8, which shows the synthesis of a 10-cycle square wave up to the first three harmonics (harmonics 1, 3, and 5).

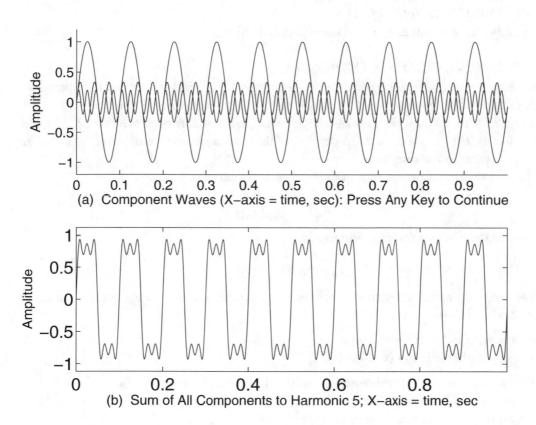

LVSynthPlotSquareSawtooth(2,10,1000)

Figure 2.8: (a) The first three weighted harmonics of a square wave; (b) Superposition of the waves shown in (a).

2.4.8   FOLDED SEQUENCE

From time to time it is necessary to reverse a sequence in time, i.e., assuming that $x[n] = [1,2,3,4]$, the folded sequence would be $x[-n]$. The operation is essentially to flip the sequence from left to right around index zero. For example, the sequence $[1,2,3,4]$ that has corresponding sample indices $[3,4,5,6]$, when folded, results in the sequence $[4,3,2,1]$ and corresponding indices $[-6,-5,-4,-3]$.

To illustrate the above ideas, we can, for example, let $x[n] = [1,2,3,4]$ with corresponding sample indices $n = [3,4,5,6]$, and compute $x[-n]$ using MathScript. We can write a simple script to accomplish the folding operation:

function [xFold,nFold] = LVFoldSeq(x,n)
xFold = fliplr(x), nFold = -fliplr(n)

For the problem at hand, we can then make the following call:

n = [3,4,5,6]; x=[1,2,3,4];
[xFold,nFold] = LVFoldSeq(x,n)
figure(6); hold on; stem(n,x,'bo'); stem(nFold,xFold,'r*')

2.4.9   EVEN AND ODD DECOMPOSITION

Any real sequence can be decomposed into two components that display even and odd symmetry about the midpoint of the sequence. A sequence that exhibits even symmetry has its first and last samples equal, its second and penultimate samples equal, and so on. A sequence that exhibits odd symmetry has its first sample equal to the negative of the last sample, its second sample equal to the negative of its penultimate sample, etc.

An even decomposition xe of a sequence x can be obtained as

$$\mathbf{xe = 0.5^*(x + fliplr(x))}$$

and the corresponding odd decomposition xo is

$$\mathbf{xo = 0.5^*(x - fliplr(x))}$$

We can write a simple function that generates the even and odd components of an input sequence $x[n]$ as follows:

function [xe,xo] = LVEvenOdd(x)
xe = (x + fliplr(x))/2; xo = (x - fliplr(x))/2;

We can illustrate use of the above script with a simple example; assuming that $x[n] = [1,3,5,7]$, we'll generate an even/odd decomposition and verify its correctness by summing the even and odd components, and comparing to the original input signal $x[n]$.

x = [1,3,5,7];
[xe,xo] = LVEvenOdd(x)
recon = xe + xo, diff = x - recon

From the above we get $xe = [4,4,4,4]$ and $xo = [-3,-1,1,3]$, the sum of which is $[1,3,5,7]$, i.e., the original sequence x.

Another useful even/odd decomposition is defined such that

$$xe[n] = xe[-n]$$

and

$$xo[n] = -xo[-n]$$

In this case, the decompositions exhibit their symmetry about $n = 0$ rather than about the midpoint of the original sequence x. For example, if $x = [1,2,3,4]$ with corresponding sample indices $n = [3,4,5,6]$, a decomposition about $n = 0$ can be accomplished by padding x with zeros in such a manner to create a new sequence with time indices symmetrical about zero. In this case, the new sequence is

$$x = [zeros(1,9),1,2,3,4]$$

having sample indices $[-6:1:6]$. The new sequence x is then decomposed as above, i.e.,

```
x = [zeros(1,9),1,2,3,4];
[xe,xo] = LVEvenOdd(x)
recon = xe + xo, diff = x - recon
```

A script that performs an even-odd decomposition about zero and returns the even and odd parts and corresponding indices without plotting is

```
function [xe,xo,m] = LVEvenOddSymmZero(x,n)
m = -fliplr(n); m1=min([m,n]); m2 = max([m,n]); m = m1:m2;
nm = n(1)-m(1); n1 = 1:1:length(n); x1 = zeros(1,length(m));
x1(n1+nm) = x; xe = 0.5*(x1 + fliplr(x1));
xo = 0.5*(x1 - fliplr(x1));
```

The call

$$[xe,xo,m] = LVEvenOddSymmZero([1,2,3],[3,4,5])$$

for example, yields

$$xe = [1.5,1,0.5,0,0,0,0,0,0.5,1,1.5]$$

$$xo = [-1.5,-1,-0.5,0,0,0,0,0,0.5,1,1.5]$$

$$m = [-5:1:5]$$

The script (see exercises below)

$$LVxEvenOddAboutZero(x, n)$$

performs a symmetric-about-zero even-odd decomposition and plots the results. Figure 2.9, which was generated by making the script call

$$LVxEvenOddAboutZero([0.9.^([0:1:30])],[0:1:30])$$

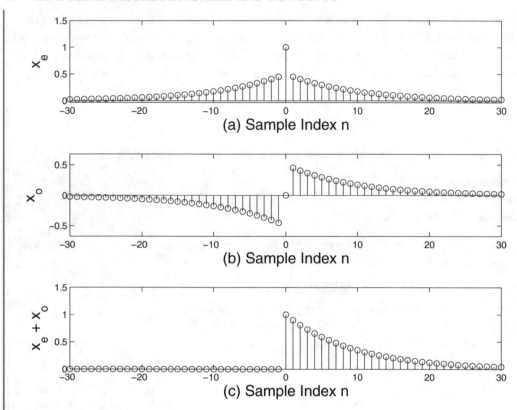

Figure 2.9: (a) Even component of a decaying exponential sequence; (b) Odd component of same; (c) Reconstruction of original exponential sequence, obtained by summing the even and odd components shown in (a) and (b).

shows the result of this process when applied to the sequence x having sample indices $n = [0{:}1{:}30]$, where

$$x = 0.9.\char`^([0{:}1{:}30])$$

2.4.10 GEOMETRIC SEQUENCE

The sum of a decreasing exponential sequence of numbers a^n, where $|a| < 1$, converges to the value $1/(1-a)$, i.e.,

$$\sum_{n=0}^{\infty} a^n \rightarrow \frac{1}{1-a} \qquad (2.2)$$

A more general statement of this proposition is that

$$\sum_{n=N}^{\infty} a^n \rightarrow \frac{a^N}{1-a} \tag{2.3}$$

which allows computation of the sum starting from a value of n greater than 0.

Another way of thinking of this is that the sum of a geometric sequence a^n (we assume $|a| < 1$) is its first term divided by one minus the convergence ratio R, where

$$R = a^n / a^{n-1}$$

For example, we can determine the sum of the following sequence using Eq. (2.3), where $N = 0$:

$$1 + 1/2 + 1/4 + 1/8 + \dots$$

Here we see that $a = 1/2$ since $a^0 = 1$, $a^1 = 1/2$, etc., so the sum is

$$\frac{1}{1 - 1/2} = \frac{1}{1/2} = 2$$

We can determine the sum of the following geometric sequence, for example,

$$1/3 + 1/9 + 1/27 + \dots$$

using Eq. (2.3) as

$$\frac{1/3}{1 - 1/3} = 1/2$$

We can verify this result with the simple MathScript call

format long; n = 1:1:50; ans = sum((1/3).^n)

which yields *ans* = 0.500000000000000.

Note that it was only necessary to use the first 50 terms of the infinite sequence to obtain a value close (in this case equal within the limitations of accuracy imposed by the computer) to the theoretical value. As a approaches unity in value, more terms are needed to obtain a sum close to the theoretical value.

Sometimes the sum of a finite number of terms of such a sequence is needed. Supposing that the sum of the first $N - 1$ terms is needed; we can subtract the sum for terms N to ∞ from the sum for all terms, i.e.,

$$\sum_{n=0}^{N-1} a^n = \sum_{n=0}^{\infty} a^n - \sum_{n=N}^{\infty} a^n = \frac{1}{1-a} - \frac{a^N}{1-a} = \frac{1 - a^N}{1-a}$$

2.4.11 RANDOM OR NOISE SEQUENCES

Noise is an ever-present background signal in communications systems. It is generated by many natural sources such as the Sun and Jupiter, lightning, many man-made sources, by active devices in electronic systems, etc. Noise assumes random values over time (rather than predictable values such as those of a sine wave, for example) which are described using statistics such as the probability density function, mean, standard deviation, etc.

It is often necessary to simulate noise in signals, and MathScript can be used to generate random sequence values using the functions

$$rand(m, n) \text{ or } randn(m, n)$$

where m and n are dimensions of the matrix of random numbers to be created.

The first function above generates a random signal having uniform distribution over the interval from 0 to 1; the second function above generates a signal having a Gaussian (or normal) distribution with a mean of 0 and standard deviation of 1 .

As an m-code example, we'll generate a signal containing noise of standard deviation 0.125 and a cosine of frequency 11 over 128 samples, and plot the result. The result from running the following m-code is shown in Fig. 2.10.

```
n = 0.125*randn(1,128); c = cos(2*pi*11*[0:127]/128);
figure(3); subplot(311); stem(c);
subplot(312); stem(n)
subplot(313); stem(n+c)
```

2.4.12 CHIRP

A sinusoid, such as a cosine wave, having a frequency that continuously increases with time, is expressed in the continuous domain as

$$y = \cos(\beta t^2)$$

Since the sampled version would have discrete sample times at nT, we would have

$$y[n] = \cos[\beta n^2 T^2]$$

Figure 2.11 shows a sampled chirp with $\beta = 49$, $T = 1/256$, and $n = 0:1:255$.

The chirp is a useful signal for testing the frequency response of a system such as a filter. A similar continuous domain technique is the use of a sweep generator to reveal the frequency response of analog circuits, such as the video intermediate frequency circuits in TV sets.

Other common uses for the chirped sinusoid are radar and ultrasonic imaging In both cases, a chirp is transmitted toward a target, with the expectation of receiving a reflection at a later time. Since the time of transmission of any frequency in the chirp is known, and the frequency and time received are known for any reflection, the difference in time between the transmission and reception

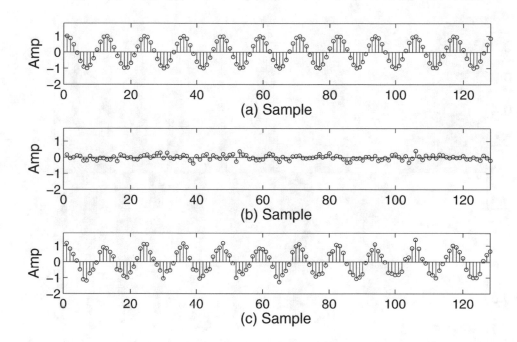

Figure 2.10: (a) A cosine sequence of amplitude 1.0; (b) Gaussian or white noise having standard deviation of 0.125; (c) The sum of the sequences at (a) and (b).

times is directly available. Since the velocity of the transmitted wave is known, the distance between the transmitter/receiver and the point of reflection on the target object can be readily determined.

MathScript's chirp function, in its simplest form, is

$$y = chirp(t, f0, t1, f1)$$

where t is a discrete time vector, $f0$ and $f1$ are the start and end frequencies, respectively, and $t1$ is the time at which frequency $f1$ occurs.

As an m-code example, we can generate a chirp that starts at frequency 0 and ends at frequency 50, over 1101 samples:

y = chirp([0:1:1100]/1100,0,1,50); figure; plot(y)

2.4.13 COMPLEX POWER SEQUENCE

While it is assumed that the reader's background encompasses complex numbers as part of a basic knowledge of continuous signals and systems, a brief summary of the common complex definitions

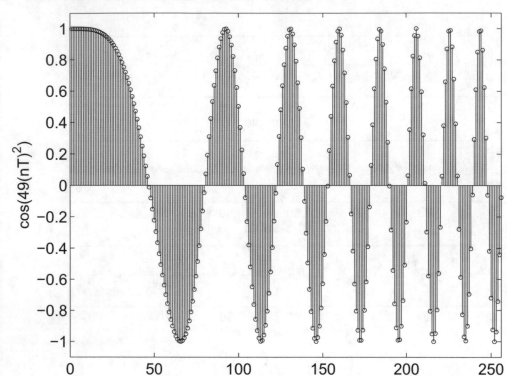

Figure 2.11: A stem plot of a sampled chirp.

and operations is found in the Appendices, which summary should provide a sufficient background for the following discussion, in which we present discrete signal sequence generation and representation using complex numbers.

An exponential of the form

$$y = e^{jx}$$

where e is the base of the natural logarithm system and j is the square root of negative one, generates a complex number lying on the unit circle in the complex plane, i.e., the number has a magnitude of 1.0. Such an exponential is equivalent to

$$e^{jx} = \cos(x) + j \sin (x)$$

As an m-code example, we can generate a complex number having a magnitude of 1.0 and lying at an angle of 45 degrees relative to the real axis of the complex plane with the following call:

y = exp(j*2*pi*(1/8))

A complex number that is repeatedly multiplied by itself generates a sequence of numbers (or samples) having real and imaginary parts which respectively define cosine and sine waves. Think of two complex numbers in polar form: the product has a magnitude equal to the product of the two magnitudes, and an angle equal to the sum of the angles. From this it can be seen that repeatedly multiplying a complex number by itself results in a sequence of complex numbers whose angles progress around the origin of the complex plane by equal increments, and the real and imaginary parts of which form, respectively, a sampled cosine sequence and a sampled sine sequence. If n represents a vector of powers, such as $0{:}1{:}N$, for example, then the complex power sequence is

$$(A\angle\theta)^n = (Ae^{j\theta})^n = A^n e^{jn\theta} = A^n(\cos n\theta + j\sin n\theta)$$

The script

$$LVxComplexPowerSeries(cn, maxPwr)$$

(see exercises below) generates a complex power sequence of the complex number cn, raised to the powers $0{:}1{:}maxPwr$.

Figure 2.12, which was created using the script just mentioned with the call

LVxComplexPowerSeries(0.99*exp(j*pi/18),40)

shows the real and imaginary parts of an entire sequence of complex numbers created by raising the original complex number W (magnitude of 0.99 at an angle of 10 degrees ($\pi/18$ radians)) to the powers 0 to 40. Note that the real part, at (c), is a cosine, and the imaginary part, at (d), is a sine wave.

Let's compute powers 0:1:3 for the complex number $[0 + j]$ and describe or characterize the resultant real and imaginary parts. The power sequence is $[j^0, j^1, j^2, j^3]$, which reduces to $[1, j, -1, -j]$, with the real parts being $[1,0,-1,0]$ and the imaginary parts being $[0,1,0,-1]$. These may be described as four-sample, single-cycle cosine and sine waves. Another way to write this would be

y = cos(2*pi*(0:1:3)/4) + j*sin(2*pi*(0:1:3)/4)

which returns the following:

y = [1,(0 +1i),(-1 + 0i),(-0 - 1i)]

Let's compute the complex power sequence W^n where $n = 0{:}1{:}4$ and $W = (\sqrt{2}/2)(1 + j)$. Note initially that $W = 1\angle45$. Then $W^{0:1:4} = [1, 1\angle45, 1\angle90, 1\angle135, 1\angle180]$, which reduces to

$$[1, 0.707(1+j), j, 0.707(-1+j), -1]$$

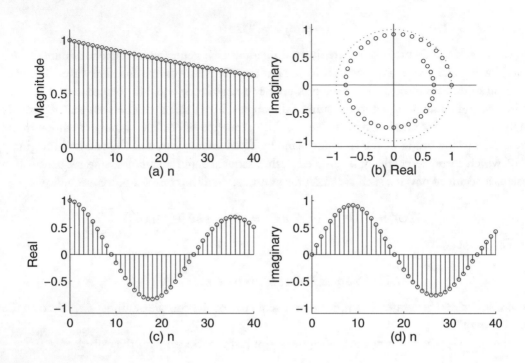

Figure 2.12: (a) Magnitude of $W = (0.99*\exp(j\pi/18))^n$ where n = 0:1:40; (b) Plot of entire power sequence in complex plane; (c) Real part of entire power sequence of W for powers 0 to 40; (d) Imaginary part of entire power sequence of W for powers 0 to 40.

To compute the expression using m-code, make the call

$$\textbf{n = 0:1:4; W = (sqrt(2)/2)*(1+j); y = W.\hat{}n}$$

As a final example, we'll compute the complex sequence values for

$$e^{-j2\pi nk/N}$$

where $n = [0,1,2,3]$, $N = 4$, and $k = 2$. This reduces to

$$\cos(\pi(0{:}1{:}3)) + j\sin(\pi(0{:}1{:}3))$$

which yields zero for all the imaginary components and for the real components we get $[1,-1,1,-1]$. This can be verified by making the call

$$\textbf{cos(pi*(0:1:3)) + j*sin(pi*(0:1:3))}$$

2.4.14 SPECIFIC FREQUENCY GENERATION

• For a given sequence length N, by choosing

$$W = M \exp(j2\pi k/N)$$

the power sequence

$$W^n = [M \exp(j2\pi k/N)]^n = M^n(\cos(2\pi nk/N) + j\sin(2\pi nk/N)) \tag{2.4}$$

will define a complex sinusoid having k cycles over N samples after each power n of W from 0 to N-1 has been evaluated. Note that the complex exponential sequence grows or decays with each succeeding sample according to the value of the magnitude M. If $|M| = 1$, the sequence has a constant, unity-amplitude; if $|M| < 1$, the sequence decays, and if $|M| > 1$, the sequence grows in amplitude with each succeeding sample.

```
function [seqCos,seqSin] = LVGenFreq(M,k,N)
% [seqCos,seqSin] = LVGenFreq(1,2,8)
n = [0:1:N-1]; arg = 2*pi*k/N;
mags = (M.^n); maxmags = max(mags);
W2n = mags.*exp(j*arg).^n;
seqCos = real(W2n); seqSin = imag(W2n);
figure(66); subplot(211); stem(seqCos);
subplot(212); stem(seqSin)
```

To illustrate use of the above script, we'll generate cosine and sine waves, having peak-to-peak amplitudes of 2 (i.e., amplitudes of unity), and having 7.5 cycles over 73 samples. A peak-to-peak amplitude of 2 means a variation in amplitude from -1 to +1, and hence an amplitude of 1.0, i.e., in the following call we set $M = 1$.

$$[\text{seqCos,seqSin}] = \text{LVGenFreq}(1,7.5,73);$$

The result is shown in Fig. 2.13.

To illustrate the generation of a growing complex exponential, we'll generate a cosine, sine pair having a frequency of 3 cycles over 240 samples, and which increases in amplitude by a factor of 1.5 per cycle. Since there are 240/3 samples per cycle, we take the 80th root of 1.5 as M. Code that makes the computation of M easier and more flexible would be

```
F=3;N=240;
[seqCos,seqSin] = LVGenFreq(1.5^(F/N),F,N)
```

The relationship given in Eq. (2.4) can be readily verified graphically and numerically using the following script:

```
function LVPowerSeriesEquiv(M,k,N)
% LVPowerSeriesEquiv(0.9,3,64)
```

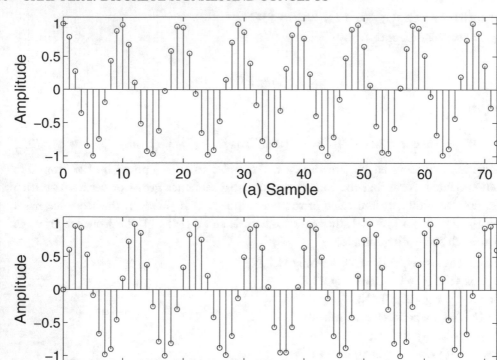

Figure 2.13: (a) The real part of a complex exponential series having 7.5 cycles per 73 samples; (b) The imaginary part of a complex exponential series having 7.5 cycles per 73 samples.

```
n = [0:1:N-1]; arg = 2*pi*k/N;
mags = (M.^n); maxmags = max(mags);
W2n = mags.*exp(j*arg).^n;
rightS = mags.*(cos(n*arg) + j*sin(n*arg));
figure(19); clf; hold on;
plot(real(W2n),imag(W2n),'bo');
plot(real(rightS),imag(rightS),'rx');
grid on; xlabel('Real'); ylabel('Imag')
axis([-maxmags,maxmags,-maxmags,maxmags])
```

Figure 2.15 shows the result from making the call

LVPowerSeriesEquiv(0.9,3,32)

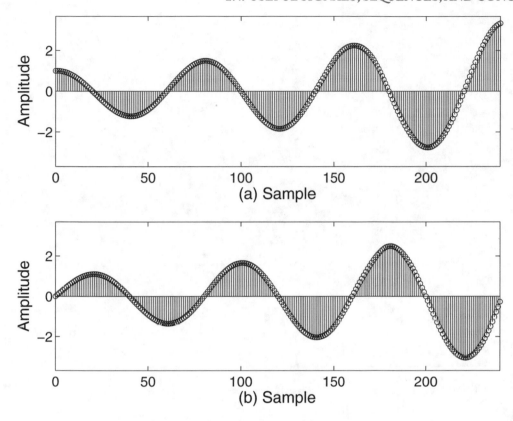

Figure 2.14: (a) The real part of a complex exponential series having one cycle per eight samples, growing in amplitude by a factor of 1.5 per cycle; (b) The imaginary part of a complex exponential series having one cycle per eight samples, growing in amplitude by a factor of 1.5 per cycle.

As a final illustration, we generate 2 cycles of a unity-amplitude cosine over 8 samples, using complex exponentials as follows

$$\textbf{real(exp(j*2*pi*2*(0:1:7)/8))}$$

and then compute the same complex exponential using the Euler identity

$$\cos(\theta) = (e^{j\theta} + e^{-j\theta})/2$$

for which a suitable call would be

$$\textbf{(exp(j*2*pi*2*(0:1:7)/8) + exp(-j*2*pi*2*(0:1:7)/8))/2}$$

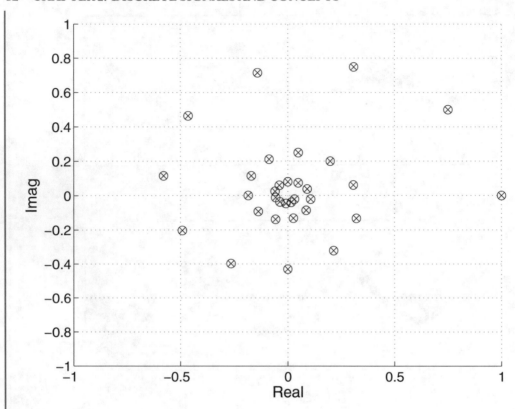

Figure 2.15: The complex power series $0.9^n \exp(\text{j} 2\pi\, 3/N)^n$ where $N = 32$ and $n = 0{:}1{:}N{-}1$, plotted as circles, and the same series, computed as $0.9^n(\cos(2\pi n3/N) + j\,\sin(2\pi n3/N))$, plotted as x's, which lie inside the circles since the two methods of computation are equivalent.

2.4.15 ENERGY OF A SIGNAL

The energy of a sequence $x[n]$ is defined as

$$E = \sum_{n=-\infty}^{\infty} x[n]x^*[n] = \sum_{n=-\infty}^{\infty} |x[n]|^2$$

where $x^*[n]$ is the complex conjugate of $x[n]$. If E is finite, $x[n]$ is called an **Energy Sequence**.

2.4.16 POWER OF A SIGNAL

The power of a signal over a number of samples is defined as

$$P = \frac{1}{(2N+1)} \sum_{n=-N}^{N} |x[n]|^2$$

A signal having finite power is called a **Power Signal**.

2.5   DISCRETE TIME SYSTEMS

2.5.1   LTI SYSTEMS

A processing system that receives an input sample sequence $x[n]$ and produces an output sequence $y[n]$ in response is called a **Discrete Time System**. If we denote a discrete time system by the operator DTS, we can then state this in symbolic form:

$$y[n] = DTS \ [x[n]]$$

A number of common signal processes and/or equivalent structures, such as FIR and IIR filtering constitute discrete time systems; they also possess two important properties, namely, 1) Time or Shift Invariance, and 2) Linearity.

A discrete time system DTS is said to be **Shift Invariant**, **Time Invariant**, or **Stationary** if, assuming that the input sequence $x[n]$ produces the output sequence $y[n]$, a shifted version of the input sequence, $x[n - s]$ produces the output sequence $y[n - s]$, for any shift of time s. Stated symbolically, this would be

$$DTS \ [x[n - s]] = y[n - s]$$

A discrete time system DTS that generates the output sequences $y_1[n]$ and $y_2[n]$ in response, respectively, to the input sequences $x_1[n]$ and $x_2[n]$ is said to be **Linear** if

$$DTS \ [ax_1[n] + bx_2[n]] = ay_1[n] + by_2[n]$$

where a and b are constants. This is called the **Principle of Superposition**.

A system that is both shift or time invariant and linear will produce the same output sequence $y[n]$ in response to the sequence $x[n]$ regardless of any shift in time of n samples. Such systems are referred to as **Linear, Time Invariant (LTI)** systems.

Example 2.1.   Demonstrate linearity and time invariance for the system below using MathScript.

$$y[n] = 2x[n]$$

We begin with code to compute $y[n] = 2x[n]$ where $x[n]$ can be scaled by the constant A. The code below generates a cosine of frequency F, scaled in amplitude by A as $x[n]$, computes $y[n]$, and then plots $x[n]$ and $y[n]$. You can change the scaling constant A and note the linear change in the output, i.e., if the input signal is scaled by A, so is the output signal (comparison of the results from running the two example calls given in the script above will demonstrate the scaling property).

```
function LVScCosine(A,N,F)
% LVScCosine(1,128,3)
% LVScCosine(2,128,3)
t = [0:1:N-1]/N; x = [A*cos(2*pi*F*t)]; y = 2*x;
subplot(2,1,1); stem(x); subplot(2,1,2); stem(y)
```

Shift invariance can be demonstrated by delaying the input signal and noting that the output is just a shifted version of the output corresponding to the undelayed input signal. For example, we can write a script similar to the above one that inserts a delay *Del* (a number of samples valued at zero) before the cosine sequence. The reader should run both of the calls given in the script below to verify the shift invariance property.

```
function LVScDelCosine(A,N,F,Del)
% LVScDelCosine(1,128,3,0)
% LVScDelCosine(1,128,3,30)
t = [0:1:N-1]/N; x = [zeros(1,Del),A*cos(2*pi*F*t)];
y = 2*x; figure(14); subplot(2,1,1);
stem(x); subplot(2,1,2); stem(y)
```

We now provide a script that will implement a simple LTI system based on scaling and delaying an input signal multiple times and adding the delayed, scaled versions together. The input argument *LTICoeff* is a row vector of coefficients, the first one of which weights the input signal, $x[n]$, the second one of which weights the input signal delayed by one sample, i.e., $x[n-1]$, and so on.

```
function [yC,nC] = LV_LTIofX(LTICoeff,x)
% [yC,nC] = LV_LTIofX([1,-2,1],cos(2*pi*12*[0:1:63]/64) )
x1 = LTICoeff(1)*x;
nC = [0:1:length(x1)-1]; yC = x1;
if length(LTICoeff)< 2
 return; end
for LTICoeffCtr = 2:1:length(LTICoeff)
xC = [LTICoeff(LTICoeffCtr)*x];
newnC = [0:1:length(xC)-1] + (LTICoeffCtr-1);
[yC, nC] = LVAddSeqs(yC,nC,xC,newnC); end
```

Example 2.2. If $x = \cos(2\pi[0:1:63]/64)$, compute $y[n]$ for the LTI system defined by

$$y[n] = LTI(x) = 2x[n] - x[n-1]$$

We make the call

[yC,nC] = LV_LTIofX([2,-1],cos(2*pi*[0:1:63]/64)); stem(nC,yC)

the results of which are shown in Fig. 2.16.

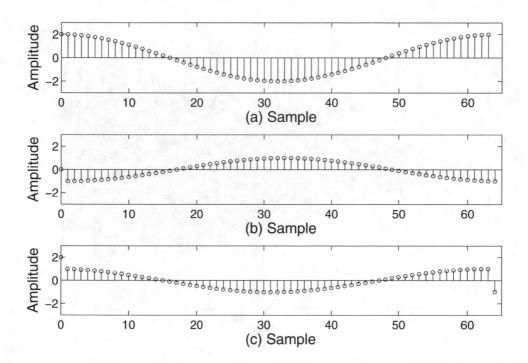

Figure 2.16: (a) The sequence $2\cos(2\pi t)$ where $t = [0:1:63]/64$; (b) The sequence $-\cos(2\pi t)$, delayed by one sample; (c) The sum or superposition of the sequences at (a) and (b).

Example 2.3.   In this example, we will see how a simple LTI system can have a frequency-selective capability. Determine the response $y[n]$ of the LTI system defined as

$$y[n] = 0.1x[n] - x[n-1] + x[n-2] - 0.1x[n-3]$$

where the test signal $x[n]$ is a linear chirp, sampled at 1000 Hz, of one second duration, that changes frequency linearly from 0 to 500 Hz.

We can use the script *LV_LTIofX* with the following call

[yC,nC] = LV_LTIofX([0.1,-1,1,-0.1],...
chirp([0:1:999]/1000,0,1,500));stem(nC,yC)

the results of which are shown in Fig. 2.17. We see that the simple four-sample LTI system has been able to process a chirp in such a way as to progressively emphasize higher frequencies. Thus this simple LTI system functions as a highpass filter.

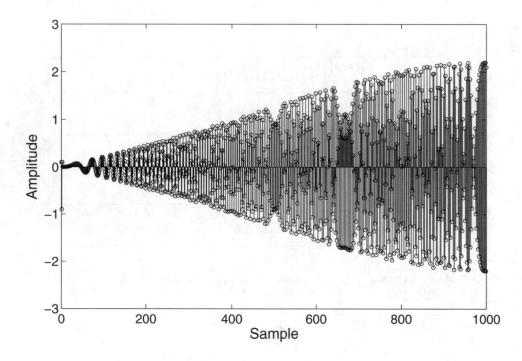

Figure 2.17: The result from convolving a linear chirp sampled at 1000 Hz of one second duration, having frequencies from 0 to 500 hz, with the LTI system defined by the coefficients [0.1,-1,1,-0.1].

The script (see exercises below)

$$LVxLinearab(a, b, f1, f2, N, Del, LTICoeff)$$

uses code similar to that above to compute $x_1[n]$, $y_1[n]$, $x_2[n]$, $y_2[n]$, $ax_1[n] + bx_2[n]$, $ay_1[n] + by_2[n]$, and LTI $[ax_1[n] + bx_2[n]]$, where LTI represents the system defined by $LTICoeff$ just as for the script LV_LTIofX. Test signal $x_1[n]$ is a cosine of frequency $f1$, and test signal $x_2[n]$ is a sine of frequency $f2$, both over N samples. An arbitrary delay of Del samples can be inserted at the leading side of $x_1[n]$ and $x_2[n]$.

Figure 2.18 was generated by making the call

LVxLinearab(2,-3,13,5,128,0,[2,-1,1,2])

which demonstrates the superposition property for the LTI system defined as

$$y[n] = 2x[n] - x[n-1] + x[n-2] + 2x[n-3] \qquad (2.5)$$

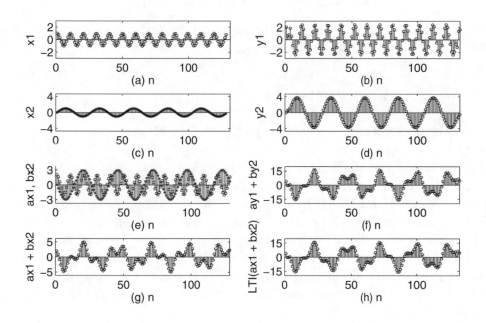

Figure 2.18: (a) $x_1[n]$; (b) $y_1[n]$; (c) $x_2[n]$; (d) $y_2[n]$; (e) $ax_1[n]$ (circles) and $bx_2[n]$ (stars); (f) $ay_1[n]$ + $by_2[n]$; (g) $ax_1[n]$ + $bx_2[n]$; (h) LTI $(ax_1[n] + bx_2[n])$.

Note that subplots (f) and (h) show, respectively, $ay_1[n] + by_2[n]$ and

$$LTI \ (ax_1[n] + bx_2[n])$$

where the LTI operator in this case represents the system Eq. (2.5). Subplots (f) and (h) show that

$$LTI \ (ax_1[n] + bx_2[n]) = ay_1[n] + by_2[n]$$

For contrast, let's consider the second order (i.e., nonlinear) system

$$y[n] = 2x^2[n] - x^2[n-1] + x^2[n-2] + 2x^2[n-3]$$

The script

$$LVxNLSabXSq(a, b, f1, f2, N, Del, NLCoeff)$$

(see exercises below) performs the superposition test on the (nonlinear) system

$$y[n] = c[0]x^2[n] + c[1]x^2[n-1] + c[2]x^2[n-2] + ...$$

where $c[n]$ are the elements of the input vector $NLCoeff$, and plots the results. The script call

LVxNLSabXSq(2,-3,13,5,128,0,[2,-1,1,2])

generated Fig. 2.19. We see from subplots (f) and (h) that $ay_1[n] + by_2[n]$ is not equal to NLS ($ax_1[n] + bx_2[n]$) where NLS represents the discrete time system

$$y[n] = \sum_{i=0}^{M} c[i]x^2[n-i]$$

with $c[i]$ defined by the input argument $NLCoeff (=[2,-1,1,2]$ in this case).

2.5.2   METHOD OF ANALYSIS OF LTI SYSTEMS

The output generated by a linear, time-invariant (LTI) may be computed by considering the input (a discrete time sequence of numbers) to be a sequence of individual sample-weighted, time-offset unit impulses. The response of any LTI system to a single unit impulse is referred to as its **Impulse Response**. The net response of the LTI system to an input sequence of sample-weighted, time-offset unit impulses is the superposition, in time offset- manner, of its individual responses to each sample-weighted unit impulse. Each individual response of the LTI system to a given input sample is its impulse response weighted by the given input sample, and offset in time according to the position of the input sample in time.

Figure 2.20 depicts this process for the three-sample signal sequence [1,-0.5,0.75], and an LTI system having the impulse response 0.7^n, where $n = 0:1:\infty$. We cannot, obviously, perform the superposition for all n (i.e., an infinite number of values), so we illustrate the process for a few values of n.

The process above, summing delayed, sample-weighted versions of the impulse response to obtain the net output, can be performed according to the following formula, where the two sequences involved are denoted $h[n]$ and $x[n]$:

$$y[k] = \sum_{n=-\infty}^{\infty} x[n]h[k-n] \tag{2.6}$$

Equation (2.6) is called the **convolution formula**.

Example 2.4.   Let $x[n] = [1, -0.5, 0.75]$ and $h[n] = 0.7^n$ Compute the first 3 output values of y as shown in subplot (h) of Fig. 2.20, using Eqn. (2.6).

We note that $x[n] = 0$ for $n < 0$ and $n > 2$. We also note that $h[n] = 0$ for values of $n < 0$. We set the range of k as 0:1:2 to compute the first three output samples, and as a result, $n = 0:1:2$,

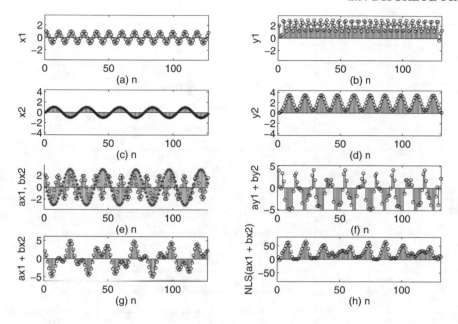

Figure 2.19: (a) $x_1[n]$; (b) $y_1[n]$; (c) $x_2[n]$; (d) $y_2[n]$; (e) $ax_1[n]$ (circles) and $bx_2[n]$ (stars); (f) $ay_1[n] + by_2[n]$; (g) $ax_1[n] + bx_2[n]$; (h) $NLS\,(ax_1[n] + bx_2[n])$.

which may be explained as follows: since the maximum value of k we will compute is 2, we need not exceed $n = 2$ since if n exceeds 2, $k - n$ is less than zero and as a result, $h[n] = 0$. To compute $y[k]$ for higher values of k, a correspondingly larger range for n is needed.

We get

$$y[0] = \sum_{n=0}^{2} x[n]h[0 - n]$$

The sum above is

$$y[0] = x[0]h[0] + x[1]h[-1] + x[2]h[-2] = x[0]h[0] = 1$$

$$y[1] = x[0]h[1] + x[1]h[0] = (1)(0.7) + (-0.5)(1) = 0.2$$

$$y[2] = x[0]h[2] + x[1]h[1] + x[2]h[0] = 0.49 + (-0.5)(0.7) + 0.75 = 0.89$$

Example 2.5. For the sequences above, compute $y[k]$ for $k = 3$.

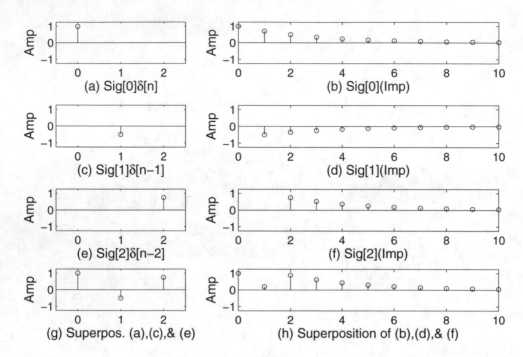

Figure 2.20: (a) First sample of signal, multiplied by $\delta(n)$; (b) Impulse response, weighted by first signal sample; (c) Second signal sample; (d) Impulse response, scaled by second signal sample, and delayed by one sample; (e) Third signal sample; (f) Impulse response scaled by third signal sample, delayed by two samples; (g) Input signal, the superposition of its components shown in (a), (c), and (e); (h) The convolution, i.e., the superposition of responses shown in (b), (d), and (f).

We set $k = 3$ and $n = 0{:}1{:}3$. Then

$$y[3] = x[0]h[3] + x[1]h[2] + x[2]h[1] + x[3]h[0]$$

which is

$$1(0.343) + (-0.5)(0.49) + 0.75(0.7) + 0(1) = 0.623$$

Example 2.6. Use m-code to compute the first 10 output values of the convolution of the two sequences $[1, -0.5, 0.75]$ and 0.7^n.

MathScript provides the function

$$conv(x, y)$$

which convolves the two sequences x and y. We make the call

$$\mathbf{conv([1\ -0.5\ 0.75],[0.7.\string^(0:1:9)])}$$

If the roles of the two sequences above had been reversed, that is to say, if we had defined the sequence $[1,-0.5, 0.75]$ as the impulse response $h[n]$ in Eq. (2.6), and $x[n] = 0.7.\string^(0:1:9)$, the result would be as shown in Fig. 2.21. Note that the first three samples of the convolution sequence are the same as shown in Fig. 2.20.

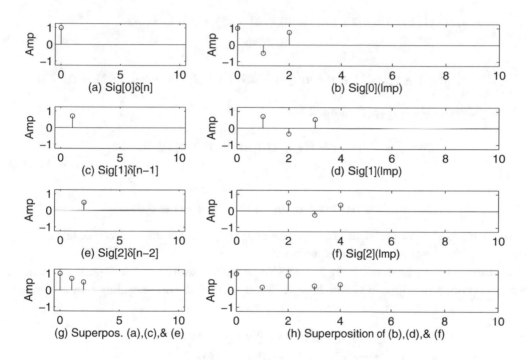

Figure 2.21: The convolution depicted in Fig. 2.20, with the roles of signal and impulse response reversed. (a) First signal sample, multiplied by $\delta\ (n)$; (b) Impulse response, scaled by first sample of signal; (c) Second sample of signal; (d) Impulse response, scaled by second signal sample, delayed by one sample; (e) Third sample of signal; (f) Impulse response scaled by third signal sample, delayed by two samples; (g) Input signal, the superposition of its components shown in (a), (c), and (e); (h) The convolution, i.e., the superposition of responses shown in (b), (d), and (f).

Example 2.7.   In this example, we'll reverse the role of signal and impulse response and show that the convolution sequence is the same. Let $x[n] = 0.7.\string^(0 : 1 : 9)$ and $h[n] = [1, -0.5, 0.75]$. Compute the first three samples of the convolution sequence.

Note that $h[n] = 0$ for $n < 0$ and $n > 2$. We set the range of k as 0:1:2, and $n = 0$:1:2. We get

$$y[0] = \sum_{n=0}^{2} x[n]h[0-n]$$

The sum above is, for $k = 0:1:2$

$$y[0] = x[0]h[0] + x[1]h[-1] + x[2]h[-2] = x[0]h[0] = 1$$

$$y[1] = x[0]h[1] + x[1]h[0] = 1(-0.5) + 0.7(1) = 0.2$$

$$y[2] = x[0]h[2] + x[1]h[1] + x[2]h[0] = 1(0.75) + 0.7(-0.5) + 0.49(1) = 0.89$$

If we additionally compute the output for $k = 3$, we get

$$y[3] = x[0]h[3] + x[1]h[2] + x[2]h[1] + x[3]h[0]$$

which yields

$$y[3] = 0 + 0.7(0.75) + 0.49(-0.5) + 0.343(1) = 0.623$$

Thus we see that the roles of the two sequences (signal and impulse response) make no difference to the resultant convolution sequence.

This can also easily be shown using MathScript's *conv* function by computing the convolution both ways and taking the difference, which proves to be zero for all corresponding output samples.

a = [1,-0.5,0.75]; b = 0.7.^(0:1:100);
c1 = conv(a,b); c2 = conv(b,a); d = c1-c2

2.5.3   GRAPHIC METHOD

An easy way to visualize and perform convolution is by time-reversing one of the two sequences and passing it through the other sequence from the left, one sample at a time. Each convolution output sample is computed by multiplying all overlapping samples and adding the products. Figures 2.22 and 2.23 show the two (equivalent) graphic orientations to compute the convolution sequence of two sequences. In Fig. 2.22, the first sequence is $0.8.^{\wedge}(0:1:9)$ and the second sequence is $0.5*[(-0.7).^{\wedge}[0:1:9]]$, and in Fig. 2.23, the roles are reversed. Inspection of subplots (c)-(f) shows that each of the four convolution values that can be computed from the illustrated overlapping sequences must be identical for the two figures since exactly the same samples from each sequence are overlapping, i.e., either sequence may be time reversed and moved through the other from left to right with the same computational result.

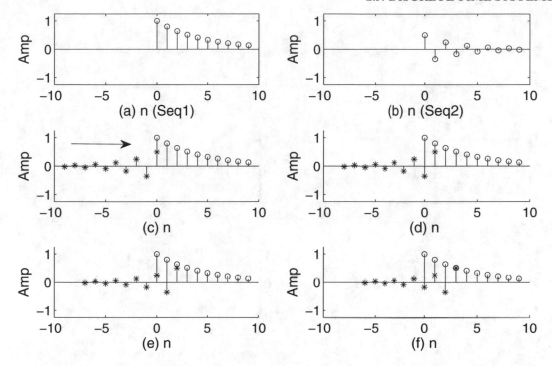

Figure 2.22: (a) First Sequence; (b) Second Sequence; (c) Second sequence time reversed (TR) and oriented to compute the first value of the convolution sequence (arrow shows direction the second sequence will slide, sample-by-sample, to perform convolution); (d) TR second sequence oriented to compute the second value of the convolution sequence; (e) TR second sequence oriented to compute the third value of the convolution sequence; (f) TR second sequence oriented to compute the fourth value of the convolution sequence.

Example 2.8.   Compute the first four values of the convolution of the sequences $[1, -1, 1, -1, 1]$ and $[1, 0.5, 0.25, 0.125]$ using the graphic visualization method.

Figure 2.24 illustrates the process. Flipping the second sequence from right to left, we get for the first convolution sequence value $(1)(1) = 1$, the second value is $(1)(-1) + (0.5)(1) = -0.5$, the third value is $(1)(1) + (0.5)(-1) + (0.25)(1) = 0.75$, and the fourth value is $(1)(-1) + (0.5)(1) + (0.25)(-1) + (0.125)(1) = -0.625$.

We can check this by making the call

$$y = \mathbf{conv}([1,0.5,0.25,0.125],[1,-1,1,-1,1])$$

which yields (with redundant zeros eliminated for brevity)

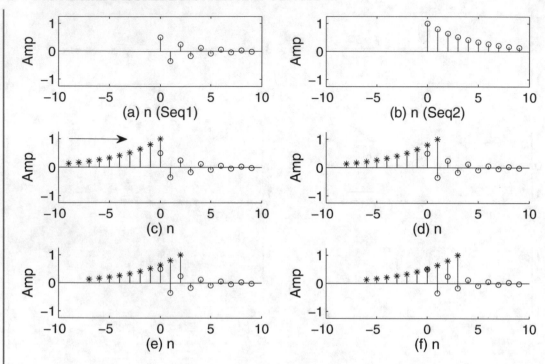

Figure 2.23: (a) First Sequence (second sequence in previous figure); (b) Second Sequence (first sequence in previous figure); (c) Second sequence time reversed (TR) and oriented to compute the first value of the convolution sequence (arrow shows direction the second sequence will slide, sample-by-sample, to perform convolution); (d) TR second sequence oriented to compute the second value of the convolution sequence; (e) TR second sequence oriented to compute the third value of the convolution sequence; (f) TR second sequence oriented to compute the fourth value of the convolution sequence.

$$y = 1, -0.5, 0.75, -0.625, 0.625, 0.375, 0.125, 0.125$$

2.5.4   A FEW PROPERTIES OF CONVOLUTION

Let's use the symbol $\circledast$ to represent convolution. Then we can compactly represent the convolution $y[n]$ of two sequences $h[n]$ and $x[n]$ as

$$y[n] = h[n] \circledast x[n]$$

Convolution is linear, so

$$y[n] = h[n] \circledast ax[n] = a(h[n] \circledast x[n])$$

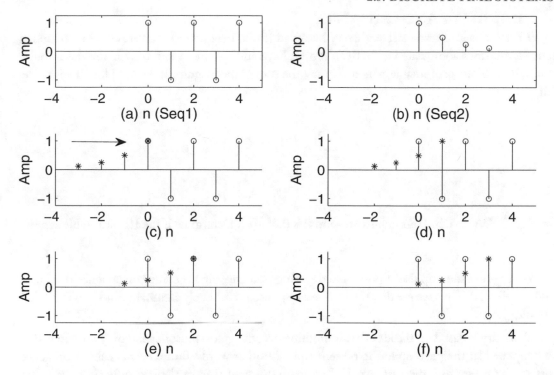

Figure 2.24: (a) First Sequence; (b) Second Sequence; (c) Second sequence time reversed (TR) and oriented to compute the first value of the convolution sequence (arrow shows direction the second sequence will slide, sample-by-sample, to perform convolution); (d) TR second sequence oriented to compute the second value of the convolution sequence; (c) TR second sequence oriented to compute the third value of the convolution sequence; (f) TR second sequence oriented to compute the fourth value of the convolution sequence.

where a is a constant.

The distributive property applies to convolution: for two sequences $x_1[n]$ and $x_2[n]$

$$h[n] \circledast (x_1[n] + x_2[n]) = h[n] \circledast x_1[n] + h[n] \circledast x_2[n]$$

The commutative property also applies, so

$$y[n] = h[n] \circledast x[n] = x[n] \circledast h[n]$$

which we showed by example above.

2.5.5   STABILITY AND CAUSALITY

An LTI system is said to be stable if every bounded (finite magnitude) input results in a bounded output (sometimes abbreviated as BIBO). An LTI system is stable if and only if the the impulse response is absolutely summable, which is to say, the sum of the magnitudes of $h[n]$ for all n is finite, i.e., if

$$\sum_{n=-\infty}^{\infty} |h[n]| < \infty$$

Example 2.9.   A certain IIR's impulse response is $0.9^n u[n]$. Determine if the IIR is a stable system.

We see that the impulse response is a decaying positive-valued geometric sequence that sums to $1/(1 - 0.9) = 10$. Thus we see that the impulse response is absolutely summable and hence the IIR is a stable system.

Causality is based on the idea that something occurring now must depend only on things that have happened in the past, up to the present time. In other words, future events cannot influence events in the present or the past. An LTI system output $y[n]$ must depend only on previous or current values of input and output. More particularly, a system is causal if

$$h[n] = 0 \text{ if } n < 0$$

Example 2.10.   Determine if the following impulse response is causal:

$$h[n] = 0.5^{n+1} u[n + 1]$$

Since the nonzero portion of the sequence begins at $n = -1$, $h[n]$ is not causal.

2.5.6   LTI SYSTEM AS A FILTER

An LTI system that has been designed to achieve a particular purpose or perform a given function, such as frequency selection or attenuation, is called a **Filter**. There are two basic types of digital filter, the FIR filter and the IIR filter, each having certain advantages and disadvantages that determine suitability for a given use.

The FIR

The Finite Impulse Response (FIR), or Transversal Filter, comprises structures or algorithms that produce output samples that are computed using only the current and previous input samples. The response of such a system to a unit impulse sequence is finite, and hence such a system is called a Finite Impulse Response filter. Figure 2.25 shows a typical example, having an arbitrary number of M delay stages, each denoted by the letter D. The cascaded delay stages act like a bucket brigade, transporting each sample, as it enters from the left, one delay stage at a time to the right. As shown, a sample sequence $s[n]$ is passing through the filter. The output of each delay stage is scaled by a multiplier according to the coefficients b_i, and all products summed yield the output. Any number of delay stages may be used, as few as one stage being possible. The larger the number of delay stages and multipliers, the greater can be the frequency selectivity.

The output of an FIR is, in general, computed using convolution. The manner of determining what an FIR's coefficients b_i should be to achieve a certain signal processing purpose is the subject of FIR design, which is covered extensively in Volume III of the series (see Chapter 1 of this volume for information on the contents of Volume III).

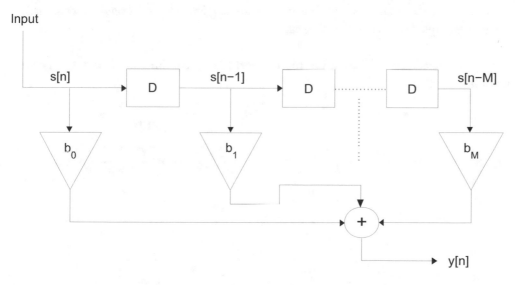

Figure 2.25: A generalized finite impulse response (FIR) filter structure. Note that the filter has M delay stages (each marked with the letter **D**) and $M + 1$ coefficient multipliers.

The IIR

The second type of basic digital filter is the **Infinite Impulse Response (IIR)** filter, which produces output samples based on the current and possibly previous input samples and previous values of its own output. This feedback process is usually referred to as a **Recursive Process**, or **Recursion**, and produces, in general, a unit impulse response which is infinite in extent. When the impulse response decays away to zero over time, the filter is stable. Figure 2.26 shows the simplest possible such filter, having one stage of delay and feedback.

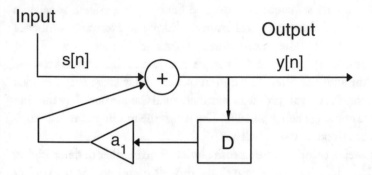

Figure 2.26: A simple recursive filter structure having a summing junction and a single feedback stage comprised of a one-sample delay element and a scaler.

A more generalized type of filter using the IIR and FIR in combination is the M-th order section, shown in Fig. 2.27.

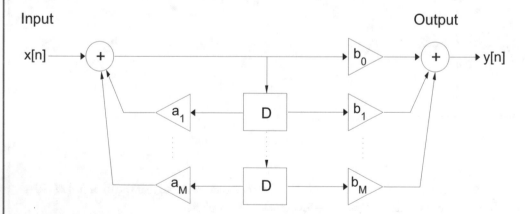

Figure 2.27: A generalized M-th order digital filter utilizing both recursive and nonrecursive computation.

Design of IIRs will be covered extensively in Volume III of this series (see Chapter 1 of this volume for a description of the contents of Volume III).

2.5.7   DIFFERENCE EQUATIONS

LTI systems can also be represented by a constant-coefficient equation that permits sequential computation of the system's output. If $x[n]$ represents an input sequence and $y[n]$ represents the output sequence of the system, an FIR can be represented by the difference equation

$$y[n] = \sum_{m=0}^{M} b_m x[n - m]$$

Example 2.11.   Compute the response to the sequence $x[n]$ = $ones(1, 4)$ of the FIR represented by the difference equation given below (assume that $x[n]$ = 0 for $n < 0$).

$$y[n] = x[n] - x[n - 1]$$

The sequence of computation is from n = 0 forward in time:

y[0] = x[0] - x[-1] = 1
y[1] = x[1] - x[0] = 1 - 1 = 0
y[2] = x[2] - x[1] = 1 - 1 = 0
y[3] = x[3] - x[2] = 1 - 1 = 0
y[4] = x[4] - x[3] = 0 -1 = -1

To do the above using MathScript, the following code is one possibility. The reader should study the code and be able to explain the purpose of or need for 1) extending x to a length of five by adding one zero-valued sample, 2) the statement y(1) = x(1), and 3) running n effectively from 1 to 5 rather than 0 to 4.

x = [ones(1,4) 0]; y(1) = x(1); for n = 2:1:5; y(n) = x(n) - x(n-1); end; ans = y

A basic difference equation for an IIR can be written as

$$y[n] = x[n] - \sum_{p=1}^{N} a_p y[n - p]$$

In this equation, the output $y[n]$ depends on the current value of the input $x[n]$ and previous values of the output $y[n]$, such as $y[n - 1]$, etc.

Example 2.12.   Compute the first four values of the impulse response of the IIR whose difference equation is

$$y[n] = x[n] + 0.9y[n - 1]$$

We use $x[n]$ = [1,0,0,0] and get

y[0] = x[0] + 0.9y[-1] = 1 + 0 = 1
y[1] = x[1] + 0.9y[0] = 0 + 0.9 = 0.9
y[2] = x[2] + 0.9y[1] = 0 + (0.9)(0.9) = 0.81
y[3] = x[3] + 0.9y[2] = 0 + (0.9)(0.81) = 0.729

A more generic form of the difference equation which includes both previous inputs and previous outputs is

$$y[n] = \sum_{m=0}^{M} b_m x[n-m] - \sum_{p=1}^{N} a_p y[n-p] \tag{2.7}$$

MathScript provides the function

$$filter(b, a, x)$$

which accepts an input vector x and evaluates the difference equation formed from coefficients b and a.

Example 2.13. A certain LTI system is defined by $b = [1]$ and $a = [1, -1.27, 0.81]$. Compute and plot the first 30 samples of the response of the system to a unit impulse, i.e., the system's impulse response.

We make the call

y = filter([1],[1, -1.27, 0.81],[1,zeros(1,29)]); figure; stem(y)

which results in Fig. 2.28.

Example 2.14. Write the difference equation corresponding to the coefficients in the previous example.

From Eqn. (2.7) we get

$$y[n] = x[n] - (-1.27y[n-1] + 0.81y[n-2]) \tag{2.8}$$

which yields

$$y[n] = x[n] + 1.27y[n-1] - 0.81y[n-2]) \tag{2.9}$$

Example 2.15. A certain filter is defined by b = [0.0466, 0.1863, 0.2795, 0.1863, 0.0466] and a = [1, −0.7821, 0.68, −0.1827, 0.0301]. Compute the filter's response to a linear chirp having frequencies varying linearly from 0 to 500 Hz.

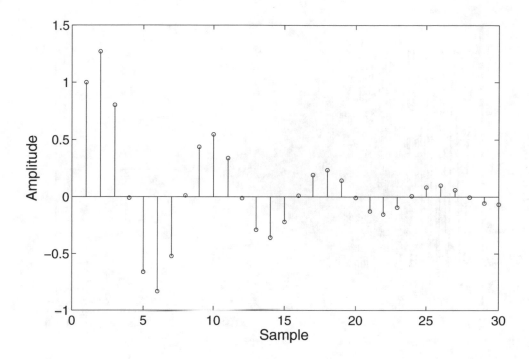

Figure 2.28: The first 30 samples of the impulse response of an LTI system defined by the coefficients $b = [1]$ and $a = [1, -1.27, 0.81]$.

The given coefficients were chosen to yield a smooth lowpass effect. The details of how to design such a filter are found in Volume III of the series (see Chapter 1 of this volume for a description of Volume III) in the chapter on Classical IIR design. Much preparation in the intervening chapters, found in this volume and the second volume of the series, will be needed to prepare the student for the study of IIR design as found in Volume III. The point of this exercise is to show how a small number of properly chosen coefficients can give a very desirable filtering result. We run the following m-code, which results in Fig. 2.29.

```
b = [ 0.0466, 0.1863, 0.2795, 0.1863, 0.0466];
a = [1, -0.7821, 0.68, -0.1827, 0.0301];
y = filter([b],[a],[chirp([0:1:999]/1000,0,1,500)]); plot(y)
xlabel('Sample'); ylabel('Amplitude')
```

2.6   REFERENCES

[1] James H. McClellan, Ronald W. Schaefer, and Mark A. Yoder, *Signal Processing First*, Pearson Prentice Hall, Upper Saddle River, New Jersey, 2003.

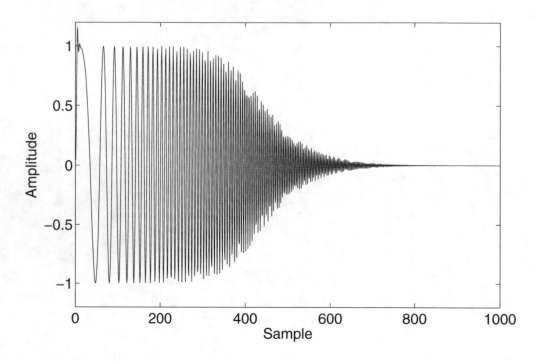

Figure 2.29: The chirp response of a filter having both *b* and *a* coefficients, chosen to yield a smooth lowpass effect.

[2] John G. Proakis and Dimitris G. Manolakis, *Digital Signal Processing, Principles, Algorithms, and Applications, Third Edition*, Prentice Hall, Upper Saddle River, New Jersey, 1996.

[3] Vinay K. Ingle and John G. Proakis, *Digital Signal Processing Using MATLAB V.4*, PWS Publishing Company, Boston, 1997.

[4] Richard G. Lyons, *Understanding Digital Signal Processing, Second Edition*, Prentice Hall, Upper Saddle River, New Jersey 2004.

2.7   EXERCISES

1. Compute and display the following sequences: 1) a sine wave having frequency = 4.5 Hz and a phase angle of 60 degrees, sampled at a rate of 2400 Hz for 2.5 seconds, beginning at time t = 0.0 second. Display the result two different ways, one way using time for the horizontal axis, and the other using sample index for the horizontal axis; 2) Repeat the above, but assume that sampling begins at t = -1.25 second rather than at t = 0.0 second.

2. Perform addition on the following pair of sequences using paper and pencil and the method of prepending and postpending zeros, then verify your answer using the script *LVAddSeqs*: 1) x1 = [1,-2,6,4], n1 = [-9,-8,-7,-6], x2 = [6,2,-3, -1], n2 = [0,1,2,3].

3. Perform addition on the following pair of sequences using the method of prepending and postpending zeros. Use m-code, and proceed by creating a figure with three subplots. In the first subplot, plot the first sequence with proper time axes; in the second subplot, plot the second sequence with the proper number of prepended zeros so that the two sequences are properly time-aligned, then plot the sum in the third subplot. Complete by verifying your answer using *AddSeqs*.

$$\sin(2\pi(5)(-1:0.01:1))$$

$$\cos(2\pi(2.75)(0:0.01:1))$$

4. Express the infinite sequence $y = 0.9^n$ (where n = 0:1:∞) as a sum of weighted unit impulse sequences.

5. Express the eight sample rectangular signal, [1,1,1,1,1,1,1,1] as a sum of unit step functions.

6. Express the sequence [0,1,2,-2,-2,-2 ...] as a sum of unit step functions, where the steady value -2 continues to $n = ∞$.

7. Compute the first 100 values of the sequence

$$y = 0.9^n u[n] - 0.9^{n-2} u[n - 2]$$

8. Compute and display eleven periods of the sequence

$$y = u[n] - u[n - 8] + \delta[n - 4]$$

9. For

$$y[n] = 0.8^{n-3} u[n - 3]$$

compute $y[-n]$ for n = -20:-1:0.

10. Let the sequence $y[n]$ = [1,2,3,4,5,6] with indices [2:1:7]. Graph the following sequences over the range n =-10:1:10.

    (a) $y[n - 1]$
    (b) $y[n + 2]$
    (c) $y[1 - n]$
    (d) $y[-3 - n]$

11. Decompose

$$y = \cos(2\pi(0:1:16)/16)$$

into even and odd components having the same length as y.

12. Decompose

$$y = \cos(2\pi(0 : 1 : 16)/16)$$

into even and odd components that are symmetrical about $n = 0$ (note that for y itself, the first sample is at $n = 0$).

13. Write a script that can generate Fig. 2.9 when called with the following statement, where the first argument is the sequence $y[n]$ and the second argument is the corresponding vector of sample indices.

LVxEvenOddAboutZero([0.9.ˆ([0:1:30])],[0:1:30])

Your script should be general enough to process a call such as

LVxEvenOddAboutZero([0.8.ˆ([4:1:30])],[-4:1:30])

14. Compute the sum of the geometric sequence 0.95^n for $n = 10$ to ∞.

15. Compute and display a signal consisting of a chirp added to noise having normal (Gaussian) distribution. The sequence should be 1024 samples long, the chirp should start at time 0.0 second and frequency 0 Hz and end at time 2.0 second at frequency 10 Hz. The noise should have a standard deviation of 1.0. Perform again for noise with standard deviations of 0.125, 0.25, 0.5, 1.0, and 2.0.

16. Write the m-code for the script

$$LVxLinearab(a, b, f1, f2, N, Del, LTICoeff)$$

as described in the text, producing the plots shown and described in Fig. 2.18, and conforming to the following function definition:

```
function LVxLinearab(a,b,f1,f2,N,Del,LTICoeff)
% Demonstrates the principle of superposition, i.e.,
% LTI(ax1 + bx2) = aLTI(x1) + bLTI(x2) where LTI is a
% linear time invariant operator defined by LTICoeff, the
% coefficients of c[n] which weight an input sequence and delayed
% versions thereof, i.e., y = LTI(x) = c[0]x[n] + c[1]x[n-1] + ..
% a and b are constants, and f1 and f2 are frequencies of cosine waves
% that are used as x1 and x2
% N is the length of the test sequences x1 and x2, and Del is a number of
% samples of delay to impose on x1 and x2 to demonstrate shift invariance.
% Test calls:
% LVxLinearab(2,5,3,5,128,0,[2])
% LVxLinearab(2,-3,13,5,128,0,[2,-1,1,2])
```

17. Write the m-code for the script

$$LVxNLSabXSq(a, b, f1, f2, N, Del, NLCoeff)$$

as described in the text and which produces the plots shown and described in Fig. 2.19.

```
function LVxNLSabXSq(a,b,f1,f2,N,Del,NLCoeff)
% Demonstrates that the principle of superposition is not true for
% a nonlinear system defined by NLSCoeff, the
% coefficients of which weight an input sequence and delayed
% versions thereof raised to the second power, i.e.,
% NLS(x) = c[0]x^2[n] + c[1]x^2[n-1] + c[2]x^2[n-2] + ...
% where c{n} are the members of the vector NLSCoeff.
% a and b are constants, and f1 and f2 are frequencies of cosine
% and sine waves, respectively, that are used as x1 and x2 in
% the superposition test i.e., does NLS(ax1 + bx2) =
% aNLS(x1)+ bNLS(x2)?
% N is the length of the test sequences x1 and x2, and Del is
% a number of samples of delay to impose on x1 and x2
% test for shift invariance.
% Test call:
% LVxNLSabXSq(2,-3,13,5,128,0,[2,-1,1,2])
```

The writing of the script can be modularized by first writing a script that will take the input coefficients $NLCoeff$ and generate the system output for a given input or test sequence $x[n]$:

```
function [yC,nC] = LVxNLSofXSq(NLCoeff,x)
% Delays, weights, and sums the square of the
% input sequence x ( = x[n] with n = 0:1:length(x)-1)
% according to yC = c(0)*x(n).^2 + c(1)*x(n-1).^ 2 + ...
% with NLCoeff (= [c[0],c[1],c[2],...]) and
% nC are the sample indices of yC.
% Test call:
% [yC,nC] = LVxNLSofXSq([1,-2,1],cos(2*pi*12*[0:1:63]/64) )
```

18. The real part of a power sequence generated from a certain complex number z having magnitude 1.0, results in 8 cycles of a cosine wave over a total of 32 samples. What is the value of z?

19. What complex number of magnitude 1.0 will generate two cycles of a complex sinusoid when raised to the power sequence 0:1:11?

20. How many cycles of a complex sinusoid are generated when the complex number

$$\cos(\pi/180) + j \sin(\pi/180)$$

is raised to the power sequence $n = 0:1:539$?

21. Write a script conforming to the following call syntax

$$LVxComplexPowerSeries(cn, maxPwr)$$

and which creates the plots shown in Fig. 2.15, where cn is a complex number which is raised to the powers 0:1:$maxPwr$.

>   **function LVxComplexPowerSeries(cn,maxPwr)**
>   **% Raises the complex number cn to the powers**
>   **% 0:1:maxPwr and plots the magnitude, the real**
>   **% part, the imaginary part, and real v. imaginary parts.**
>   **% Test calls:**
>   **% LVxComplexPowerSeries(0.69*(1 + j),50)**
>   **% LVxComplexPowerSeries(0.99*exp(j*pi/18),40)**

22. Determine if the following difference equations represent stable LTI systems or not:

$$y[n] = x[n] + y[n-1]$$

$$y[n] = x[n] + 1.05y[n-1]$$

$$y[n] = x[n] + 0.95y[n-1]$$

$$y[n] = x[n] + 1.2y[n-2]$$

$$y[n] = x[n] - 1.2y[n-2]$$

$$y[n] = x[n] - 1.8y[n-1] - 0.8y[n-2]$$

$$y[n] = x[n] - 1.8y[n-1] + 0.8y[n-2]$$

$$y[n] = x[n] + 1.27y[n-1] - 0.81y[n-2]$$

$$y[n] = x[n] - 1.27y[n-1] + 0.81y[n-2]$$

23. Compute and plot the response of the following systems ((a) through (e) below) to each of the following three signals:

$$x[n] = u[n] - u[n-32]$$

and

$$x(n) = [1, \text{zeros}(1,100)]$$

and

$$x(n) = \text{chirp}([0:1/1000:1],0,1,500)$$

Be sure to review how to convert between b and a coefficients suitable for a call to the function $filter$ and the coefficients in a difference equation. Note particularly Eqs. (2.8) and (2.9). For situations where the system is an FIR, you can also use the script LV_LTIofX.

a) The system defined by the difference equation

$$y[n] = x[n] + x[n-2]$$

b) The system defined by the difference equation

$$y[n] = x[n] - 0.95y[n-2]$$

c) The system defined by the coefficients $a = [1]$ and

$$b = [0.1667, 0.5, 0.5, 0.1667]$$

d) The system defined by the coefficients $b = [1]$ and

$$a = [1, 0, 0.3333]$$

e) The system defined by the coefficients

$$b = [0.1667, 0.5, 0.5, 0.1667]$$

$$a = [1, 0, 0.3333]$$

24. Use paper and pencil and the graphical method to compute the first five values of the convolution sequence of the following sequence pairs, then check your answers by using the MathScript function $conv$.

    (a) **[(-1).^(0:1:7)], [0.5*ones(1,10)]**
    (b) **[0.1,0.7,1,0.7,0.1], [(-1).^(0:1:9)]**
    (c) **[1,1], [(-0.9*j).^(0:1:7)]**
    (d) **[(exp(j*pi)).^(0:1:9)], [ones(1,3)]**

25. Verify that the commutative property of convolution holds true for the argument pairs given below, using the script LV_LTIofX, and then, for each argument pair, repeat the exercise using the MathScript function $conv$. Plot results for comparison.

(a) chirp([0:1/99:1],0,1,50) and [1,1];
(b) chirp([0:1/99:1],0,1,50) and [1,0,1];
(c) [1,0,-1] and chirp([0:1/999:1],0,1,500)
(d) [1,0,-1] and chirp([0:1/999:1],-500,1,500)

26. Does a linear system of the form

$$y = kx + c$$

where k and c are constants, and x is an independent variable obey the law of superposition, i.e., is it true that

$$y(ax_1 + bx_2) = ay(x_1) + by(x_2)$$

where a and b are constants? Prove your answer.

27. Sound travelling from a point of origin in a room to a listening point typically takes many transmission paths, including a direct path, and a number of reflected paths. We wish to determine how the sound quality at the listening point is affected by the multiple transmission paths, i.e., assuming that the source emits a linear chirp of unity amplitude covering all the frequencies of interest to us, what will the frequency response at the listening point be? Assume that the transmission paths from the source point to the listening point can be modeled as an LTI system, and that we can sample a microphone output at the listening point and plot the received signal over time. Since the chirp frequency increases linearly over time, a plot of received amplitude versus time is equivalent to one of received amplitude versus frequency.

We can simulate this experiment by knowing the various path lengths and the speed of sound, which allows us to compute the sound transit times for the direct and reflected paths. Assume that the hypothetical microphone output used at the listening point is sampled at 10 kHz, and the speed of sound in air is 1080 ft/sec., and that we can model the transfer function with the direct and two reflected paths. To perform the simulation, follow these steps:

1) Compute all path lengths in terms of time, and since frequency response is not dependent on bulk delay, only on the time (or phase) differences between the interfering signals, the transit time of the direct (i.e., shortest) path can be subtracted from all transit times.

2) Specify each of the reflected (i.e., delayed) paths as a number of equivalent samples of delay, based on the sample rate at the microphone. Then we can specify the system as a set of delays and appropriate amplitudes, and use the function LV_LTIofX to compute the output.

3) A typical call to LV_LTIofX should be of the form

$$[yC, nC] = LV\_LTIofX(LTICoeff, x)$$

where x is a linear chirp over a duration of one second, having 10,000 samples and chirping from 0 to 5 kHz, and the input argument $LTICoeff$ is of the form

[1,zeros(1,n1-1),RG1,zeros(1,(n2-n1)-1),RG2]

where $n1$ is the number of samples of delay between the time of arrival (TOA) of the direct wave and the TOA of the first reflected wave, $(n2 - n1)$ is the additional delay from the TOA of the first reflected wave to the TOA of the second reflected wave, $RG1$ is the gain of the first reflected path relative to that of the direct path, which is defined as 1.0, and $RG2$ is the gain of the second reflected path relative to that of the direct path.

The following test parameters consist of three path lengths in feet and the relative amplitudes of each wave upon arrival at the listening point. For each set of test parameters, determine and make the appropriate call to LV_LTIofX, and plot the output yC versus frequency.

(a) Path Lens = [2.16, 2.268, 2.376]; RelAmps = [1,0.98,0.96];
(b) Path Lens = [8.64, 8.748, 8.856]; RelAmps = [1,0.98,0.96];
(c) Path Lens = [10.8, 11.664, 14.148]; RelAmps = [1,0.96,0.88];
(d) Path Lens = [75.6, 86.076, 98.388]; RelAmps = [1,0.93,0.81];
(e) Path Lens = [54, 54.108, 54.216]; RelAmps = [1,1,1];
(f) Path Lens = [54, 54.108, 54.216]; RelAmps = [1,0,1]; (1st refl path damped)

Now repeat the exercise, but instead of using a simple linear chirp, use the following complex chirp:

t = [0:1/9999:1]; fMx = 5000;
x = chirp(t,0,1,fMx) + j*chirp(t,0,1,fMx,'linear',90);

Also, instead of using the script LV_LTIofX, use the function $filter$, with input argument $a = 1$ and b equal to the input argument $LTICoeff$ as would have been used with LV_LTIofX. Since the result is complex, plot the absolute value of the result of the filtering operation.

Figure 2.30 shows the result from this alternate procedure, using the parameters specified at (c) above. Figure 2.31 (resulting from the call at (f) above) shows what happens when the first reflected path is severely attenuated, as perhaps by an acoustic absorber inserted into the path to alter the frequency response at the listening point.

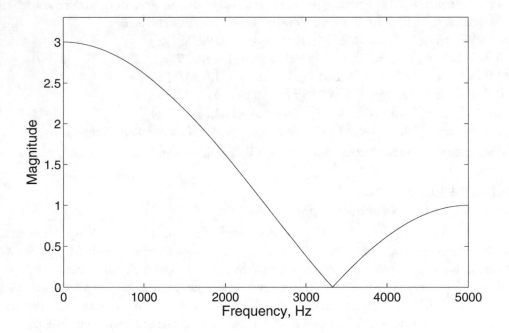

Figure 2.30: An estimate of the frequency response between source and listening points in a room, modeled with a direct and two reflected paths. The first reflected path is only one sample longer in duration than the direct path. The second reflected path is only one sample of delay longer than the first reflected path. Both reflected waves arrive at the same amplitude as that of the direct wave.

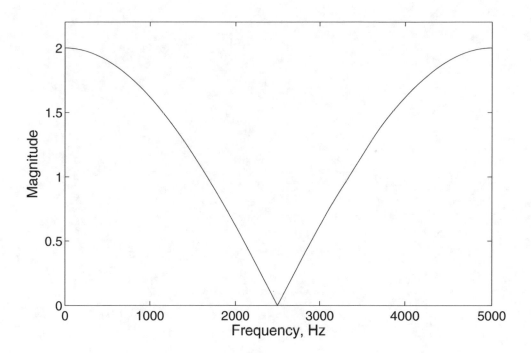

Figure 2.31: An estimate of the frequency response between source and listening points in a room, modeled with a direct and two reflected paths. The first reflected path is one sample longer in duration than the direct path, but has been hypothetically severely attenuated (to amplitude zero) with an acoustic absorber to observe the effect on frequency response. The second reflected path is only one sample of delay longer than the first reflected path. The second reflected wave arrives at the same amplitude as that of the direct wave.

CHAPTER 3

Sampling and Binary Representation

3.1 OVERVIEW

Having equipped ourselves with the knowledge of a number of common signal types and the basic concepts of LTI systems in the previous chapter, we continue our study of DSP with the conversion of an analog signal from the continuous domain to the discrete or sampled domain, and back again. Real-world analog signals (such as audio and video signals) first enter the digital realm via the process of sampling. We begin our discussion with two very important requirements of sampling, namely, the minimum acceptable or **Nyquist** sample rate and the need for bandlimiting a signal prior to sampling. With these all-important principles established, we discuss normalized frequency, which is the basis for evaluating and describing frequency content and response in the digital domain. Nyquist rate and normalized frequency are among the most fundamental of concepts associated with digital signal processing, and little further discussion of the topic can be meaningfully had until the reader understands them well.

In this chapter, in addition to the Nyquist rate and normalized frequency, we also discuss the basics of binary counting and formats and analog-to-digital and digital-to-analog conversion. These are important topics since they refer to implementation, which is always with limited precision representation of numbers. The theory of digital signal processing, such as digital filtering, the DFT, etc., is usually taught as though all numbers are of perfect accuracy or infinite precision. When digital signal processing algorithms are implemented on a computer, for example, all numbers are stored and all computations are made with only finite precision, which can adversely affect results. Much of the DSP literature concerns these issues, and thus familiarity with the principles and nomenclature of conversion and quantization is essential.

3.2 SOFTWARE FOR USE WITH THIS BOOK

The software files needed for use with this book (consisting of m-code (.m) files, VI files (.vi), and related support files) are available for download from the following website:

http://www.morganclaypool.com/page/isen

The entire software package should be stored in a single folder on the user's computer, and the full file name of the folder must be placed on the MATLAB or LabVIEW search path in accordance with the instructions provided by the respective software vendor (in case you have encountered this notice before, which is repeated for convenience in each chapter of the book, the software download only needs to be done once, as files for the entire series of four volumes are all contained in the one downloadable folder).

See Appendix A for more information.

3.3   ALIASING

In 1928, Nyquist, working at the Bell Telephone Laboratories, discovered that in order to adequately reconstruct a sinusoid, it was only necessary to obtain two samples of each cycle of the sinusoid. So if we have a continuous-valued voltage representing a single frequency sinusoid, we need to obtain amplitude samples of the signal twice per cycle. If we sample regularly at equal intervals, we can describe the sampling operation as operating at a certain frequency, and obviously this frequency, F_S, will have to be at least twice the frequency of the sinusoid we are sampling.

- If a sinusoid is sampled fewer than two times per cycle, a phenomenon called **Aliasing** will occur, and the sampled signal cannot be properly reconstructed. When aliasing occurs, a signal's original, pre-sampling frequency generally appears in the sampler output as a different apparent, or aliased, frequency.

- In signals containing many frequencies, the sampling rate must be at least twice the highest frequency in the signal–this ensures that each frequency in the signal will be sampled at least twice per cycle.

The preceding statement leads to the question, "how do you know what the highest frequency is in the signal you are quantizing?" The general answer is that the only way to know is to completely control the situation by filtering the analog signal before you sample it; you would use an analog (continuous domain) lowpass filter with a cutoff frequency at half the sampling frequency. This ensures that in fact the sampling rate is more than twice the highest frequency in the signal.

Such an analog filter is called an **Anti-Aliasing Filter**, or sometimes, simply an **Aliasing Filter**. The usual arrangement is shown in Fig. 3.1; the time domain signal, which might have unlimited or unknown bandwidth, passes through an anti-aliasing filter in which all frequencies above one-half the sampling rate are removed. From there, the actual sampling operation is performed: a switch is momentarily closed and the instantaneous amplitude of the signal is stored or held on a capacitor while the unity-gain-buffered output of the Sample-and-Hold is fed to an Analog-to-Digital Converter (ADC), an example of which we'll cover in detail after completing our discussion of aliasing.

Figure 3.2 shows what happens when an 8 Hz sine wave is sampled at a rate of 9 Hz rather than the minimum acceptable value of 16 Hz–the output sequence looks just like the output sequence that would have been generated from sampling a 1 Hz-inverted-phase-sine wave at a rate of 9 Hz.

- An aliased frequency in a digital sequence has forever lost its original identity. There are, assuming an unlimited bandwidth frequency being input to a sampler without an anti-aliasing

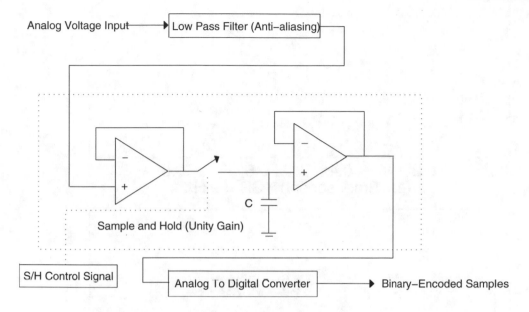

Figure 3.1: A typical sampling arrangement, showing an anti-aliasing filter followed by a sample-and-hold, which captures and holds an analog value to be digitized by the ADC.

filter, literally an infinite number of frequencies which could be aliased into even a single frequency lying below the Nyquist rate (half the sampling rate). In this situation, once a frequency becomes aliased, there is no way to reverse the aliasing process and determine which of the infinite number of possible source frequencies was in fact the original frequency.

- Samples are like snapshots or individual frames in a motion picture of action which is continuous in nature. If we take 30 snapshots or frames per second of human actors, we are certain that we have a good idea of everything that takes place in the scene simply because human beings cannot move fast enough to do anything of significance in the time interval between any two adjacent snapshots or frames (note that the human eye itself takes snapshots or samples of visually received information and sends them one after another to the brain). Imagine then if we were to lower the frame rate from one frame (or snapshot) every 30th of a second to, say, one frame every five seconds. At this low rate, it is clear that the actors could engage in a huge range of activities and complete them between frames. From looking at the sequence of frames, each one five seconds apart, we could not say what had taken place between frames.

Likewise, when sampling a waveform, too low a sampling rate for the frequencies present in the waveform causes loss of vital information as to what went on between samples.

Example 3.1. Illustrate the effect of sampling an 8 Hz sinusoid at sample rates varying from 240 Hz downward to 8 Hz.

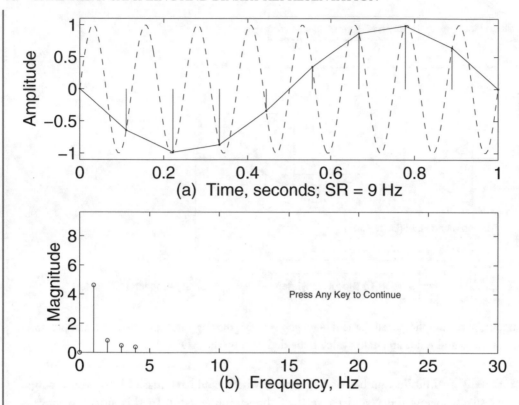

Figure 3.2: (a) Eight cycles of a sine wave, sampled at only 9 Hz; in order to avoid aliasing, at least 16 samples should have been taken. Due to the phenomenon of aliasing, the nine samples give the impression that the original signal was actually a one cycle sine wave. (b) A frequency content plot, obtained using the Discrete Fourier Transform, confirms what the eye sees in (a).

The script

LVAliasingMovieSine

performs a demonstration of the effect of various sampling rates applied to eight cycles of a sinusoid, partly automated and partly requiring user manual input. A similar demonstration is given by the VI

DemoAliasingMovieSineVI

which requires on-screen manual variation of the sample rate F_S of an 8 Hz sine wave between F_S = 120 Hz and F_S = 8 Hz.

Figure 3.3, plot (a), shows eight cycles of a sine wave in which the sample rate is 240, or about 30 samples per cycle.

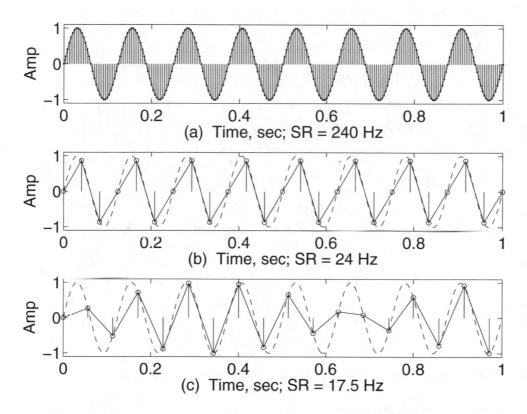

Figure 3.3: (a) Eight cycles of a sinc wave taken over a 1 second period, with a stem plot of 240 samples thereof superimposed; (b) The same eight-cycle sine, sampled at a rate of 24 Hz; (c) Sampled at 17.5 Hz.

If instead of a 240 Hz sampling rate, we use 24 Hz instead, we can still see what appears to be eight cycles of a waveform, since there are three samples per cycle (at 0, 120, and 240 degrees).

At a sample rate of 17.5 Hz, it is still possible to see eight cycles. When the sample rate falls to exactly 16 Hz (Fig. 3.4, plot (a), exactly two samples per cycle), we have the unfortunate situation that the sampling operation was synchronized with the wave at phase zero, yielding only samples of zero amplitude, and resulting in the spectrum having no apparent content.

When the sample rate falls to 14 Hz, a careful counting of the number of apparent cycles yields an apparent 6 Hz wave, not 8 Hz, as shown in Fig. 3.4, plot (b). At a sampling rate of 9 Hz, we see the samples outlining a perfect inverted-phase 1 Hz wave!

You may have noticed some apparent relationship between input frequency, sampling rate, and apparent output frequency. For example, with an input frequency of 8 Hz and a sampling rate of 9 Hz, we saw an apparent output frequency of 1 Hz, with the sine wave's phase inverted.

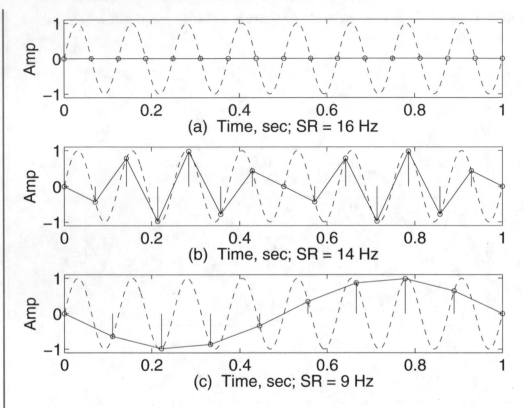

Figure 3.4: (a) An 8 Hz sine wave sampled at 16 Hz; (b) Sampled at 14 Hz; (c) Sampled at 9 Hz.

3.4   FOLDING DIAGRAM

- For any given sampling rate, a **Folding Diagram** that shows the periodic nature of the sampling function can be constructed. Such a diagram allows easy determination of the apparent output frequency of the sampler for a given input frequency. The sampling function, as illustrated by a folding diagram, "folds" input frequencies around odd multiplies of half the sampling rate.

Example 3.2.   Construct a Folding Diagram for a 9 Hz sampling rate and determine the output frequency and phase if an 8 Hz signal is input to the sampler.

Figure 3.5, plot (a), shows a standard Folding Diagram using a sampling rate of 9 Hz. Use is self-explanatory, except for the phase of the output signal compared to the input signal. When the slope of the folding diagram is positive, the output frequency is in-phase with the input, and when the slope of the folding diagram is negative, the output frequency's phase is inverted when compared to the input signal's phase.

Thus we see in Fig. 3.5, plot (a), a sampling rate of 9 Hz, an input frequency of 8 Hz, and an apparent output frequency of 1 Hz, which is phase reversed since the input frequency lies between one-half the sampling rate and the sampling rate, i.e., the folding diagram has a negative slope for an input frequency of 8 Hz.

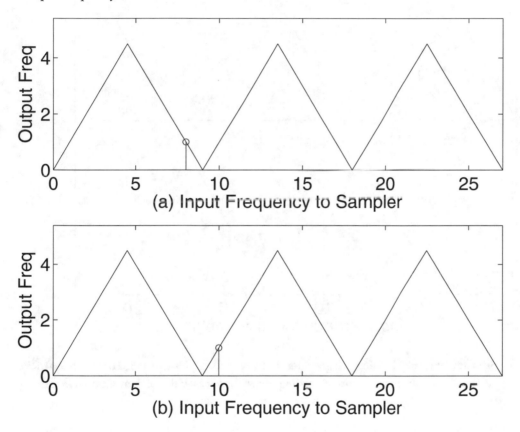

Figure 3.5: (a) The Frequency Folding Diagram for a sampling rate of 9 Hz, with an 8 Hz input signal to the sampler; (b) Folding Diagram for a 9 Hz sampling rate, with a 10 Hz input signal.

You can see in Fig. 3.5, plot (b), an input frequency *greater* than the sampling rate by, say, 1 Hz, still shows up in the output with an apparent frequency of 1 Hz, according the Folding Diagram for a 9 Hz sampling rate. In this case, however, its phase will be the same as that of the input signal at 10 Hz.

An easier way to see the phase reversal is to use the Folding Diagram shown in Fig. 3.6, in which the "downside" areas of input frequency that result in phase reversal are graphed such that the output frequency has a negative sign, indicating phase reversal.

In this diagram, if the input frequency is a multiple of half the sampling frequency, the output frequency sign is indeterminate. The actual output frequency will be equal to the input frequency,

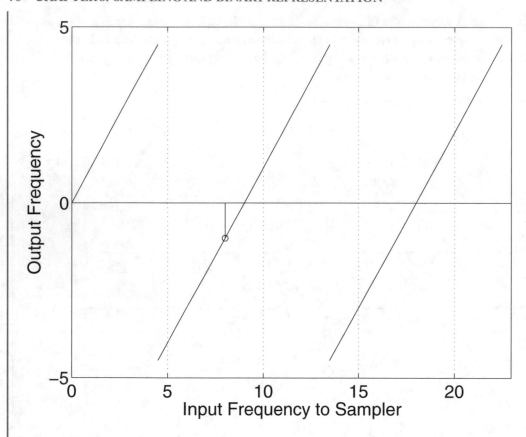

Figure 3.6: Bipolar Frequency Folding Diagram for 9 Hz, depicting phase reversed output frequencies as having negative values, thus making correct reading easier than it is with the conventional Frequency Folding Diagram.

with a variable amplitude, except for the case in which the input frequency bears that special phase relationship (as seen in Fig. 3.4), plot (a), in which case all samples have a uniform value of zero, conveying no frequency information.

To see the effect of aliasing, you can run the script

$$LVAliasing(SR, Freq)$$

and try various frequencies *Freq* with a given sampling rate *SR* to verify the correctness of the Folding Diagram.

The script evaluates the expression

sin(2*pi*(0:1/SR:1-1/SR)*Freq)

for user-selected values of SR (Sample Rate) and $Freq$ (sinusoid frequency).

You can start out with the call

LVAliasing(100,2)

which results in Fig. 3.7, plot (a). A sample rate of 100 Hz with a frequency of 102 Hz is shown in plot (b). Note that the two plots are identical. The result would also have been identical if the frequency had been 202 Hz, or 502 Hz, etc. Rerun *LVAliasing* with the sampling rate as 100 Hz, and the frequency as 98 Hz. The result is shown in Fig. 3.7, plot (c)—an apparent 2 Hz sine wave, but phase inverted, as would have been predicted by a folding diagram based on a 100 Hz sample rate.

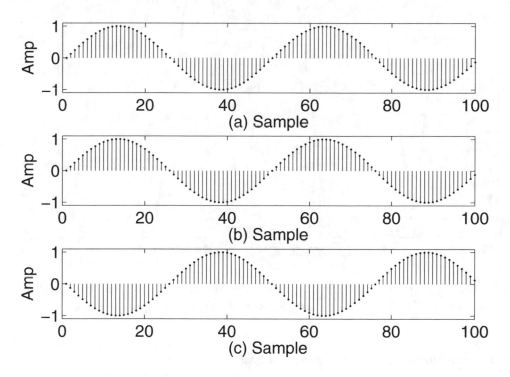

Figure 3.7: (a) A sine wave with a frequency of 2 Hz sampled at 100 Hz; (b) A sine wave with a frequency of 102 Hz sampled at 100 Hz; (c) A sine wave with a frequency of 98 Hz sampled at 100 Hz.

Example 3.3. Construct a folding diagram for a sampling rate of 100 Hz, and plot input frequencies at 8 Hz, 108 Hz, 208 Hz, 308 Hz, 408 Hz, and 508 Hz.

Figure 3.8 shows the result; note that all input frequencies map to the same apparent output frequency, 8 Hz (marked with a horizontal dotted line). Note further that only input frequencies lying between 0 and 50 Hz map to themselves as output frequencies. All input frequencies higher than 50 Hz appear in the output as aliased frequencies.

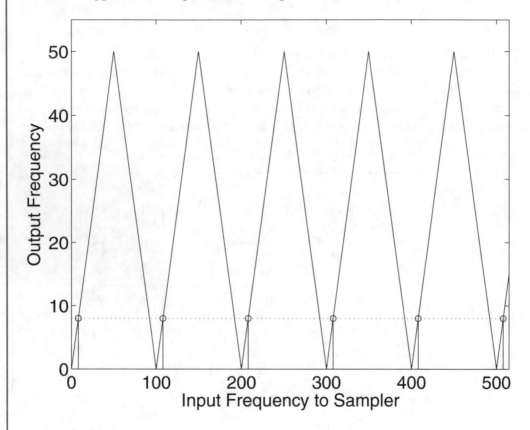

Figure 3.8: Folding diagram for 100 Hz sample rate with sinusoidal inputs to the sampler of 8, 108, 208, 308 408, and 508 Hz, showing net output frequency of 8 Hz. for all input frequencies.

Example 3.4.   Demonstrate aliasing in an audio signal.

The script

$$ML\_AliasingChirpAudio(SR, StartFreq, EndFreq)$$

has been provided to both illustrate and make audible aliasing in an audio signal. A variable sample rate SR, and the lowest frequency $StartFreq$ and highest frequency $EndFreq$ of a linear chirp are specified as the input arguments.

Figure 3.9, plot (a) shows the result of the call

ML_AliasingChirpAudio(3000,0,1500)

which specifies a lower chirp limit *StartFreq* of 0 Hz and an upper chirp limit *EndFreq* of 1500 Hz, which is right at the Nyquist rate of 3000/2 = 1500 Hz. You can see a smooth frequency increase in the spectrogram in plot (b). Frequency 1.0 represents half the sampling frequency (3 kHz), or 1500 Hz in this case. If you have a sound card on your computer, the chirp should sound automatically when the call above is made.

- To use the script *ML_AliasingChirpAudio* with LabVIEW, restrict the value of *SR* to one of the following values: 8000, 11025, 22050, 44100.

In Fig. 3.9, plot (c), the lower chirp limit is 0 Hz, and the upper chirp limit is 3000 Hz, with the same 3000 Hz sampling rate. You can see a smooth frequency increase in the spectrogram (plot (d)) up to the midpoint in time, and then the apparent frequency smoothly decreases to the starting frequency. Listening to the chirp frequency go up and then suddenly reverse should fix in your mind what happens in aliasing.

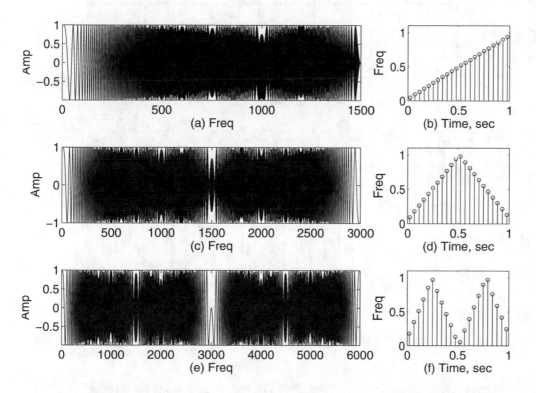

Figure 3.9: (a) Chirp from 0 Hz to 1500 Hz, sampled at 3000 Hz; (b) Spectrogram (frequency versus time) of (a); (c) Chirp from 0 Hz to 3000 Hz, sampled at 3000 Hz; (d) Spectrogram of (c); (e) Chirp from 0 Hz to 6000 Hz, sampled at 3000 Hz; (f) Spectrogram of (e).

Running the experiment one more time, let's use 6000 Hz as the upper chirp frequency, with the same sampling frequency of 3000 Hz. The result is shown in plot (e), with the spectrogram in plot (f).

3.5   NORMALIZED FREQUENCY

Consider a 16-Hz cosine wave sampled at 32 Hz for one second, with proper anti-aliasing. The resulting sequence would look like Fig. 3.10, plot (a). On the other hand, a 32-Hz cosine sampled at 64 Hz results in the same apparent signal, an alternation between +1 and -1, as shown in Fig. 3.10, plot (b).

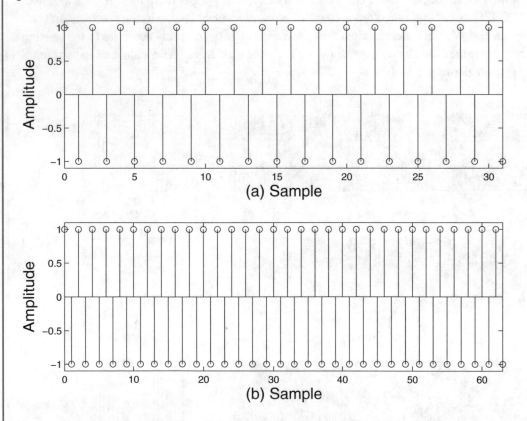

Figure 3.10: (a) A cosine wave of 16 Hz sampled at 32 Hz; (b) A cosine wave of 32 Hz sampled at 64 Hz.

If the sample rate had been (for example) 2048 Hz, the Nyquist limit would have been 1024 Hz, and a 1024 Hz cosine in the original analog signal would show up as an alternation with a two-sample period. If instead, the cosine's frequency had been one-half the Nyquist limit (in this case 512 Hz), there would have been four samples per cycle, and so on.

- **Two sinusoids, sampled at different rates, that bear the same frequency relative to their respective Nyquist rates, produce sample sequences having the same frequency content.**

- **Two sinusoids, sampled at different rates, that bear the same frequency and phase relative to their respective Nyquist rates, produce essentially the same sample sequences.**

Example 3.5. Verify the above statements using MathScript.

The m-code

N=32; FrN2=0.5; t=0:1/N:1-1/N;
figure; stem(cos(2*pi*t*FrN2*(N/2)))

allows you to verify this by holding *FrN2* (a fraction of the Nyquist rate $N/2$, such as 0.5, etc.) constant, while changing the sample rate N. In the above call, the phase angle (0) remains the same as N changes, and the resulting sequences will be identical except for total number of samples. Note that *FrN2* = 1 generates the Nyquist rate.

That the two sinusoids must have the same phase to produce apparently identical sequences may be observed by modifying the above call by inserting a phase angle in the cosine argument, and changing it from one sampling operation to the next. For example, run this m-code

pa = 1.5; N = 32; fN2 = 0.25; t = 0:1/N:1-1/N;
figure; stem(cos(2*pi*t*fN2*N/2+pa))

and note the resulting sample sequence. Then follow with the call below and note the result.

pa = 1; N = 64; fN2 = 0.25; t = 0:1/N:1-1/N;
figure; stem(cos(pi*t*fN2*N+pa))

Had the value of *pa* (phase angle) been the same in both experiments above, so would have been the resulting sequences. The sequences produced have the same apparent frequency content but different sample values (and hence different appearances when graphed) due to the phase difference.

We define **Normalized Frequency** as the original signal frequency divided by the Nyquist rate.

Definition 3.6.

$$F_{norm} = F_{orig} / F_{Nyquist}$$

Once an analog signal has been sampled, the resulting number sequence is divorced from real time, and the only way to reconstruct the original signal properly is to send the samples to a DAC at the original sampling rate. Thus it is critical to know what the original sampling rate was for purposes of reconstruction as a real time signal.

When dealing with just the sequence itself, it is natural to speak of the frequency components relative to the Nyquist limit.

- It is important to understand the concept of normalized frequency since the behavior of sequences in digital filters is based not on original signal frequency (for that is not ascertainable from any information contained in the sequence itself), but on normalized frequency.

Not only is it standard in digital signal processing to express frequencies as a fraction of the Nyquist limit, it is also typical to express normalized frequencies in radians. For example, letting $k = 0$ in the expression

$$W^n = [\exp(j2\pi k/N)]^n$$

yields the complex exponential with zero frequency, which is composed of a cosine of constant amplitude 1.0 and a sine of constant amplitude 0.0. If N is even, letting $k = N/2$ yields the net radian argument of π (180 degrees) and the complex exponential series

$$W^n = [\exp(j\pi)]^n = (-1)^n$$

which is a cosine wave at the Nyquist limit frequency.

Thus we see that radian argument 0 generates a complex exponential with frequency 0, the radian argument π generates the Nyquist limit frequency, and radian arguments between 0 and π generate proportional frequencies therebetween. For example, the radian argument $\pi/2$ yields the complex exponential having a frequency one-half that of the Nyquist limit, that is to say, one cycle every four samples. The radian argument $\pi/4$ yields a frequency one-quarter of the Nyquist limit or one cycle every eight samples, and so forth.

Hence the normalized frequency 1.0 (the Nyquist limit) may be taken as a short form of the radian argument 1.0 times π, the normalized frequency of 0.5 represents the radian argument $\pi/2$, and so forth.

- Radian arguments between 0 and π radians correspond to normalized frequencies between 0 and 1, i.e., between DC and the Nyquist limit frequency (half the sampling rate).

Example 3.7.   A sequence of length eight samples has within it one cycle of a sine wave. What is the normalized frequency of the sine wave for the given sequence length?

A single-cycle length of two samples represents the Nyquist limit frequency for any sequence. For a length-eight sequence, a single cycle sinusoid would therefore be at one-quarter of the Nyquist frequency. Since Nyquist is represented by the radian argument π, the correct radian frequency is $\pi/4$ or 0.25π, and the normalized frequency is 0.25.

Example 3.8.   A sequence of length nine samples has within it 1.77 cycles of a cosine wave. What is the normalized frequency of the cosine wave?

The Nyquist frequency is one-half the sequence length or 4.5; the normalized frequency is therefore 1.77/4.5 = 0.3933. We can construct and display the actual wave with the call

$$\text{figure; stem(cos(2*pi*1.77*(0:1:8)/9))}$$

Example 3.9. Generate, in a sequence length of 9, a sine wave of normalized frequency 0.5 (or radian frequency of 0.5π).

A plot of the signal can be obtained by making the call

$$\text{figure; stem(sin(2*pi*(0:1:8)/9*(0.5*4.5)))}$$

The frequency argument for the call is constructed as the product of the normalized frequency 0.5 and the Nyquist limit frequency, which is 9/2 = 4.5 for this example.

Example 3.10. Generate the complex exponential having a normalized frequency of 0.333π in a length-eleven sequence, and plot the imaginary part.

The following is a suitable call:

$$\text{figure; stem(imag(exp(j*2*pi*(0:1:10)/11*(0.333*5.5))))}$$

• The VI

DemoComplexPowerSeriesVI

allows you to use the mouse to move the cursor in the complex plane and to view the corresponding complex power series. As you move the cursor, the complex number corresponding to the cursor position is used as the kernel of a complex power series, the real and imaginary parts of which are displayed in real time. Figure 3.11 shows an example of the VI in use.

Example 3.11. Use the VI DemoComplexPowerSeriesVI to generate and display the real and imaginary parts of a complex power sequence generated from a complex number having a magnitude of 0.9 and a normalized frequency of 0.25.

A normalized frequency of 0.25 implies a frequency of one-quarter of the Nyquist rate. The angle is thus $\pi/4$ radians, yielding the complex number 0.707 + j0.707. After scaling by the magnitude of 0.9, we get the complex number 0.636 + j0.636, or

$$0.9\text{*exp(j*pi*0.25)}$$

To see the power sequence generated from this number, use the VI

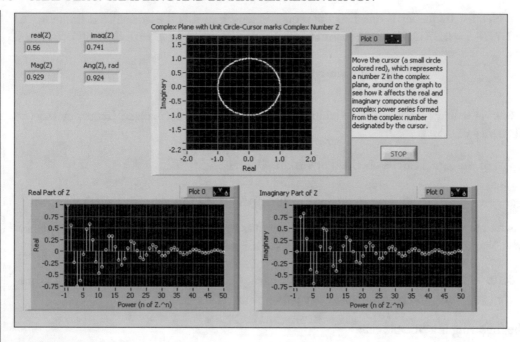

Figure 3.11: A VI that demonstrates the relation between a complex number and the power sequence arising from it. The complex number designated by the cursor position in the complex plane is used as the kernel of a power sequence (the first 50 powers are computed), and the real and imaginary parts are displayed, updated in real time as the user drags the cursor in the complex plane.

DemoComplexPowerSeriesVI

and adjust the cursor to obtain a position yielding magnitude 0.9 and normalized frequency 0.25, or alternatively, make the MathScript call

LVxComplexPowerSeries(0.636*(1+j), 50)

which displays the same information for the complex number 0.636*(1+j) , computed and displayed for the first 50 powers (creation of this script was part of the exercises of the previous chapter).

Figure 3.12 shows the display when the cursor is set at complex coordinates (0.636 + j*0.636). The resultant power sequence is a decaying complex sinusoid exhibiting one cycle over eight samples, i.e., one-quarter the frequency of the Nyquist limit.

Suitable m-code calls to plot the real and imaginary parts of a complex power sequence are, for the example at hand,

figure; stem(real((0.9*exp(j*pi*0.25)). ^(0:1:29)))

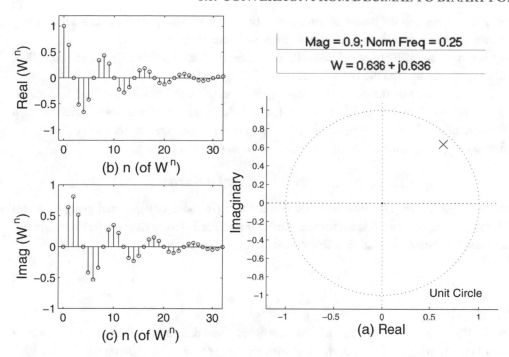

Figure 3.12: (a) Complex plane, showing the complex number having magnitude 0.9 and angle 45 degrees, i.e., 0.636 + j0.636; (b) Real part of complex power sequence; (c) Imaginary part of complex power sequence.

and

$$\text{figure; stem(imag((0.9*exp(j*pi*0.25)). ^(0:1:29)))}$$

You can experiment by changing any of the magnitude, normalized frequency, or length of the exponent vector.

- A script that allows you to dynamically see the impulse response and frequency response generated by a pole or zero, or pair of either, as you move the cursor in the complex plane is

ML_DragPoleZero

3.6   CONVERSION FROM DECIMAL TO BINARY FORMAT

Digital systems represent all signals as sequences of numbers, and those numbers are expressed in binary format using only two symbols, 1 and 0. Thus prior to discussing analog-to-digital conversion, it is necessary to understand the basics of binary counting, since the output of an ADC is a binary number that represents a quantized version of the analog input sample.

In our standard counting system, "Base 10," the number 10 and its powers, such as 100, 1,000, 10,000, and so forth are used as the basis for all numerical operations.

For example, if you saw the decimal number "3,247" in print and read it out loud, you might say "three thousand, two hundred and forty-seven, i.e., three thousand (or three times 10 to the third power), plus another two hundred (two times 10 to the second power), plus another forty (four times 10 to the first power), plus another seven (seven times 10 to the zero power).

$$3247 = 3 \cdot 10^3 + 2 \cdot 10^2 + 4 \cdot 10^1 + 7 \cdot 10^0$$

Computers use a Base 2 arithmetic system—numbers are coded or expressed using only two symbols, 1 and 0, which serve as weights for the various powers of 2. To see how the decimal number 9 can be expressed in binary format, let's first construct a few powers of 2:

$2^0 = 1$
$2^1 = 2$
$2^2 = 4$
$2^3 = 8$

To make the conversion from decimal to binary, we must determine whether each power of 2 should be given the weight of 1 or 0, starting with the largest power of 2 being used. Let's look ahead to the answer, which is [1 0 0 1] and note that

$$9 = 1 \cdot 2^3 + 0 \cdot 2^2 + 0 \cdot 2^1 + 1 \cdot 2^0 \tag{3.1}$$

In Eq. (3.1), we call the weighting values, 1 or 0, for each power of 2, **Binary Digits**, or more commonly **Bits**. In any binary number, the rightmost bit (the weight for 2^0) is called the **Least Significant Bit**, or **LSB**, while the leftmost bit is called the **Most Significant Bit**, or **MSB**.

To arrive at (3.1) algorithmically, we'll use a simple method called **Successive Approximation,** which is commonly used to convert analog or continuous domain sample values to binarily-quantized sample values, in which we consider whether or not each bit, starting with the MSB, should ultimately be given a weight of 1 or 0. An example should make the method clear.

Example 3.12.   Convert decimal number 9 to binary notation using the Method of Successive Approximation.

We start by setting the MSB, 2^3 (= 8), to 1. Since $1 \cdot 2^3$ is less than 9, we retain the value 1 for the 2^3 bit. Proceeding to 2^2 (= 4), if we add 4 to the 8 we have (8 = $1 \cdot 2^3$), we get 12 (= $1 \cdot 2^3$ + $1 \cdot 2^2$). But 12 exceeds the number to be converted, 9, so we reset the weight of 2^2 from 1 to 0. Proceeding to 2^1 (= 2), we see that the 8 we have ($1 \cdot 2^3 + 0 \cdot 2^2$) plus another 2 would result in a sum total of 10, so we reset the 2^2 weight to 0. Proceeding to 2^0 (= 1), we see that adding 1 to the current sum total of 8 gives the required 9, so we keep the 2^0 weight at 1. Thus we have 9 = [1 0 0 1] in binary notation.

3.7   QUANTIZATION ERROR

Note that so far we have been examining the conversion of integer values between decimal and binary formats. In general, the amplitudes of a sequence of samples will not always be integer multiples of the LSB, and thus there will be, in general, a difference between the quantized value and the original value. This difference is called the **Quantization Error** and may be made arbitrarily small by using a larger number of quantization bits. We will examine this in detail below after the presentation of a script to perform analog-to-digital conversion on a sequence and compute the quantization error.

For purposes of illustrating the general principles of ADC and DAC, we will continue using this integer format, i.e., quantizing signals as integral multiples of the LSB. There are, of course, binary formats that contain fractional parts as well as an integer part. See reference [1] for further details.

3.8   BINARY-TO-DECIMAL VIA ALGORITHM

Note that part of the process of successive approximation involves evaluating a test binary word's equivalent decimal value, i.e., a digital-to-analog step, which is the reverse of the overall procedure being conducted. Prior to presenting a program to convert an entire sequence of decimal numbers (or equivalently, from the realm of real-world signals, a sequence of analog or continuous-valued samples) to binary format, we'll describe several simple scripts to convert a binary representation back to decimal.

Example 3.13.   Write a simple program that will convert a binary number to decimal form.

First, we receive an arbitrary length binary number bn, and then note that $bn(1)$ is the weight for $2^{LenBN-1}$, $bn(2)$ is the weight for $2^{LenBN-2}$, and so forth, where $LenBN$ is the length of bn. We then initialize the converted decimal number dn at 0. The loop then multiplies each bit by its corresponding power of two, and accumulates the sum.

```
function [dn] = LVBinary2Decimal(bn)
% [dn] = LVBinary2Decimal([1,0,1,0,1,0,1,0])
Lenbn = length(bn); dn = 0; for ctr = Lenbn:-1:1;
dn = dn + 2^(ctr-1)*bn(Lenbn-ctr+1); end;
```

A simplified (vectorized) equivalent of the above code is

```
function [dn] = LVBinary2DecimalVec(bn)
% [dn] = LVBinary2DecimalVec([1,0,1,0,1,0,1,0])
dn = sum(2.^(length(bn)-1:-1:0).*bn);
```

Yet another variation uses the inner or dot product of the power of two vector and the transpose (into a column vector) of the binary number:

```
function [dn] = LVBin2DecDotProd(bn)
% [dn] = LVBin2DecDotProd([1,0,1,0,1,0,1,0])
```

dn = 2.^(length(bn)-1:-1:0)*bn';

3.9   DECIMAL-TO-BINARY VIA ALGORITHM

Now we are ready, in the following example, to present a script for algorithmic conversion of a decimal number (zero or positive) to binary format.

Example 3.14.   Write a script that will quantize an input decimal-valued sample to a given number of bits, and print out descriptions of each step taken during the successive approximation process.

```
function [BinOut, Err] = LVSuccessAppSingle(DecNum,Bits)
% [BinMat, Err] = LVSuccessAppSingle(5.75,4)
if length(DecNum)>1; error('DecNum must be scalar')
end
BinOut = zeros(1,Bits); BinWord = zeros(1,Bits);
for ctr = 1:1:Bits
Status = (['Setting Bit ',num2str(ctr),' to 1'])
BinWord(1,ctr) = 1
DecEquivCurrBinWord = LVBinary2DecimalVec(BinWord)
Status = (['Subtracting decimal equiv of BinWord from sample being quantized'])
diff = DecNum - DecEquivCurrBinWord
if diff < 0;
Status = (['Resetting Bit ',num2str(ctr),' to 0'])
BinWord(1,ctr) = 0;
end
end
Status = (['Final Binary Word:'])
BinOut(1,:) = BinWord;
Err = DecNum - LVBinary2DecimalVec(BinWord);
```

The following printout occurs after making the call

$$[BinMat] = LVSuccessAppSingle(1.26,2)$$

```
Status = Setting Bit 1 to 1
BinWord = 1 0
DecEquivCurrBinWord = 2
Status = Subtracting decimal equiv of BinWord from sample being quantized
diff = -0.7400
Status = Resetting Bit 1 to 0
Status = Setting Bit 2 to 1
```

BinWord = 0 1
DecEquivCurrBinWord = 1
Status = Subtracting decimal equiv of BinWord from sample being quantized
diff = 0.2600
Status = Final Binary Word:
BinMat = 0 1
Err = 0.2600

3.10   OFFSET TO INPUT TO REDUCE ERROR

The method of successive approximation as described above always chooses bit weights which add up to a quantized value which is less than or equal to the actual value of the input signal. As a result, the approximation of the quantized signal to the actual signal is biased an average of one-half LSB toward zero, and the maximum quantization error is 1.0 times the LSB.

Example 3.15.   Demonstrate that a bias to the input signal of 0.5 LSB away from zero results in a maximum quantization error of one-half LSB rather than 1.0 LSB.

We present a script that can convert an entire sequence of decimal numbers to binary equivalents, quantized to a specified number of bits, and where a bias of one-half LSB to the input signal can be optionally added. This script is not the most efficient one possible; in the exercises below, the reader is presented with the task of writing a script that will conduct the successive approximation technique simultaneously on an entire input sequence, thus eliminating the outer loop, which can make the computation very slow for long sequences.

```
function [BinMat,Err] = LVSuccessApp(DecNum,MaxBits,LSBBias)
% [BinMat,Err] = LVSuccessApp(7.5*[sin(0.5*pi*[0:1/18:1])]+7.5,4,1);
BinMat = zeros(length(DecNum),MaxBits);
DecOutMat = zeros(1,length(DecNum));
if LSBBias==0; xDecNum = DecNum;
else; xDecNum = DecNum + 0.5; end
for DecNumCtr = 1:1:length(DecNum)
BinTstWord = zeros(1,MaxBits);
for ctr = 1:1:MaxBits; BinTstWord(1,ctr) = 1;
DecEquivCurrBinTstWord = LVBinary2DecimalVec(BinTstWord);
diff = xDecNum(DecNumCtr) - DecEquivCurrBinTstWord;
if diff < 0; BinTstWord(1,ctr) = 0;
elseif diff==0; break; else; end
end
BinMat(DecNumCtr,:) = BinTstWord;
DecOutMat(1,DecNumCtr) = LVBinary2DecimalVec(BinTstWord);
```

```
Err = DecNum-DecOutMat;
end
figure(78); clf; subplot(211); ldn = length(DecNum);
xvec = 0:1:ldn-1; hold on; plot(xvec,DecNum,'b:');
plot(xvec,DecNum,'bo'); stairs(xvec,DecOutMat,'r');
xlabel('(a) Sample'); ylabel('Amplitude')
subplot(212); stairs(xvec,Err,'k');
xlabel('(b) Sample'); ylabel('Error')
```

The call

$$[BinMat,Err] = LVSuccessApp(7.5*[sin(0.5*pi*[0:1/18:1])]+7.5,4,0);$$

results in Fig. 3.13, in which the quantized values are plotted in stairstep fashion rather than as discrete points. Note that the error is positive, and has a maximum amplitude of about 1.0, i.e., one LSB.

The following call specifies that prior to conversion, 0.5LSB (i.e., 0.5 in this case since the LSB is 1.0) be added to the signal:

$$[BinMat,Err] = LVSuccessApp(7.5*[sin(0.5*pi*[0:1/18:1])]+7.5,4,1);$$

The result is shown in Fig. 3.14, in which it can be seen that the error is both positive and negative, but does not exceed 0.5, i.e., one-half LSB.

3.11   CLIPPING

The call

$$[BinMat] = LVSuccessApp(10*[sin(2*pi*[0:1/18:1])]+10,4,1);$$

results in Fig. 3.15. As only four bits were specified for the quantization, we can only represent the numbers 0-15, whereas the input waveform ranges in value from 0 to 20. The algorithm outputs its maximum value of 15 (binary [1,1,1,1]) for signal values of 14.5 and higher; the output is said to be clipped since graphically it appears that its upper portions (samples 2 through (7) have simply been cut or clipped off.

The situation above can be remedied by increasing the number of bits from 4 to 5. We make the call

$$[BinMat] = LVSuccessApp(10*[sin(2*pi*[0:1/18:1])]+10,5,1);$$

which results in Fig. 3.16. For five bits, the possible output values range from 0 to 31, easily encompassing the signal at hand.

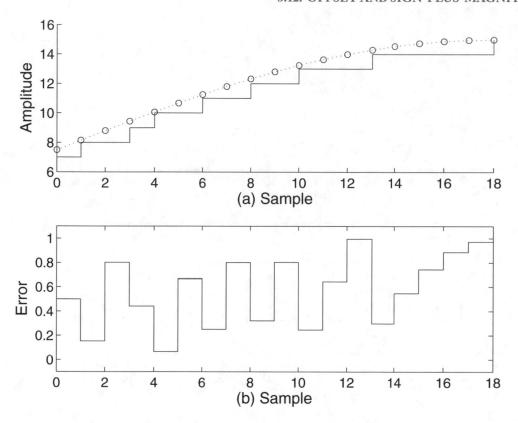

Figure 3.13: (a) Analog signal (dashed), analog sample values (circles), and quantized samples (stairstep); (b) Quantization error, showing a maximum magnitude equal to the LSB.

3.12   OFFSET AND SIGN-PLUS-MAGNITUDE

The method just discussed assumes all numbers to be converted are zero or positive. This is a common method of digitization, the **Offset Method** (sometimes called the **Unipolar** method), in which a typical analog signal having both positive and negative voltage values is given a DC offset so that it is entirely nonnegative. Another format is the **Sign Plus Magnitude Method**, in which the most significant bit represents the sign of the number, and the remaining bits represent the magnitude.

Example 3.16.   Assume that a binary number is in the sign-plus-magnitude format, with a sign bit value of 0 meaning positive and 1 meaning negative. Write a simple script to convert such a number to decimal equivalent.

Initially, we set aside the first bit of the binary number, which is the sign bit, and convert the remaining bits to decimal as before. If the sign bit has the value 1, the answer is multiplied by -1, otherwise, it is left positive.

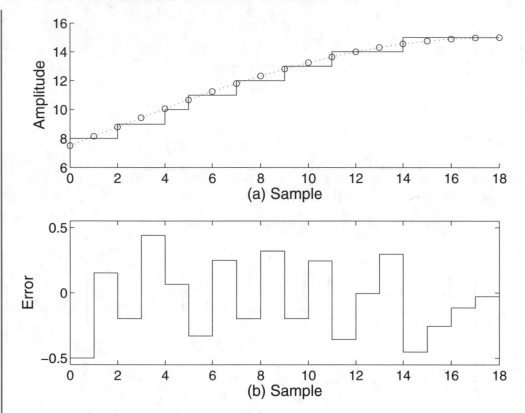

Figure 3.14: (a) Analog signal (dashed), analog sample values (circles), and quantized samples (stairstep); (b) Quantization error, showing a maximum magnitude error of 0.5 LSB.

```
function [dn] = LVSignPlusMag2Dec(bn)
% [dn] = LVSignPlusMag2Dec([1,0,1,0,1])
pbn = fliplr(bn(2:length(bn)));
dn = (-1)^(bn(1))*sum(2.^(0:1:length(pbn)-1).*pbn)
```

The script

$$BinOut = LVxBinaryCodeMethods(BitsQ, SR, Bias, Freq, Amp, CM, PlotType)$$

(see exercises below) affords the opportunity to experiment with the analog-to-digital conversion of a test sine wave, using either of the two formats mentioned above. It allows you to select the number of quantization bits $BitsQ$, the sample rate SR, amount of bias to the input signal $Bias$, test sine wave frequency $Freq$, test sine wave amplitude Amp, coding method CM, which includes *Sign plus*

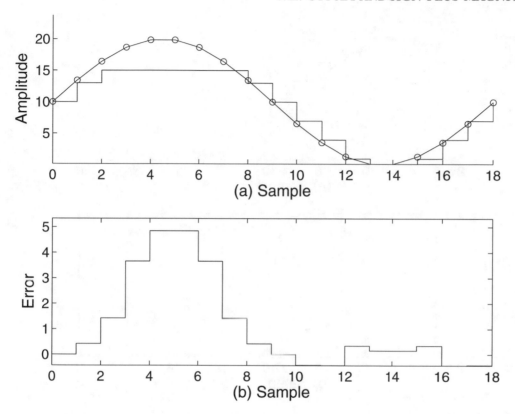

Figure 3.15: (a) Analog signal (dashed), analog sample values (circles), and quantized samples (stairstep). Due to clipping, the sample values for sample times 2-6 are limited to quantized level 15; (b) Error signal, difference between original signal and the quantized signal at sample times.

Magnitude and *Offset*, and *PlotType*, which displays the quantized output in volts or as multiples of the LSB.

Example 3.17. Demonstrate the Sign Plus Magnitude conversion method (or format) using the script *LVxBinaryCodeMethods*.

Figure 3.17, plots (a) and (b), show a portion of the result from using parameters of *SR* = 1000 Hz, *Freq* = 10 Hz, *Amp* = 170, *BitsQ* = 2 Bits, *Bias* = None, and *CM* = *Sign plus Mag*, and *PlotType* as volts. The call is

BinOut = LVxBinaryCodeMethods(2,1000,0,10,170,1,1)

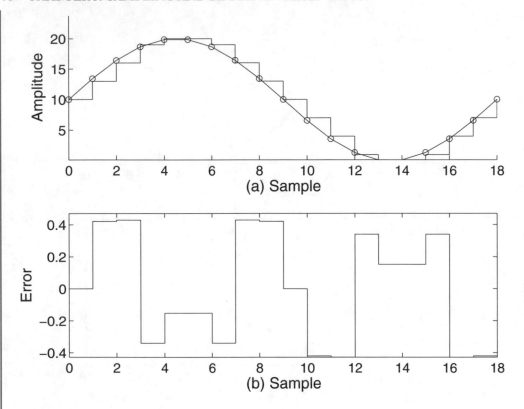

Figure 3.16: (a) Analog signal (dashed), analog sample values (circles), and quantized samples (stairstep); (b) Quantization error, showing relief from the clipping of the previous example, achieved by increasing the number of quantization bits, allowing for quantization up to decimal equivalent 31 ($2^5 - 1$).

Plot (a) of the figure shows the analog voltage and the discrete, quantized samples; plot (b) shows the noise voltage, which is the difference between the original analog signal and the quantized version.

The use of only two bits, no input bias, and sign plus magnitude coding results in a quantized signal which is almost completely noise (albeit in this case the noise component is highly periodic in nature). As can be seen in plot (a), only two samples are quantized at or near their true (original analog) voltages; all other quantized voltages are zero, resulting in a noise signal of very high amplitude relative to the original signal.

The situation can be considerably improved by biasing the input signal by one-half LSB away from zero. Plots (c) and (d) show the result.

Example 3.18.   Demonstrate the Offset conversion method using the script *LVxBinaryCodeMethods*.

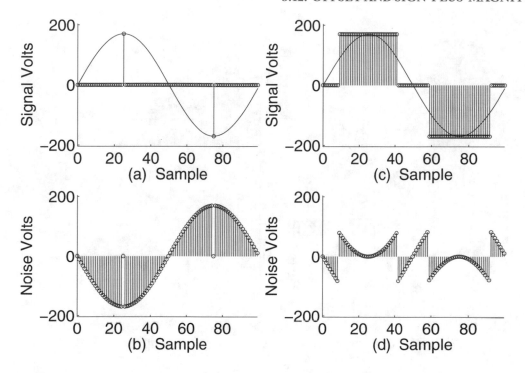

Figure 3.17: (a) A portion of an analog or continuous 10 Hz sine wave (solid); quantized using 2 bits (one sign and one magnitude), LSB = 170 volts, stem plot; (b) Quantization Noise; (c) Analog 10 Hz sine wave, solid; quantized using 2 bits (one sign and one magnitude), input biased one-half LSB away from zero for both positive and negative values, LSB = 170 volts; (d) Quantization Noise.

The offset method gives 2^n quantization levels, as opposed to the sign plus magnitude coding method, which yields only $2^n - 1$ distinct levels. Using *SR* = 1000 Hz, *Freq* = 10 Hz, *Amp* = 170, *BitsQ* = 2 Bits, *Bias* = One-Half LSB, and *CM* = Offset, and *PlotType* as volts, we get Fig. 3.18, plots (a) and (b). The call is

BinOut = LVxBinaryCodeMethods(2,1000,1,10,170,2,1)

In the previous figure, the quantized output was depicted in volts; the actual output of an ADC is a binary number giving the sample's amplitude as a multiple of the value of the LSB. In Fig. 3.18, plots (c) and (d), the quantized output and the quantization noise are both plotted as multiples of the value of the LSB, which is 113.333 volts, computed by dividing the peak amplitude (340 volts) of the input signal by $(2^2 - 1)$, which is the number of increments above zero needed to reach the peak amplitude from the minimum amplitude, which is set at zero in accordance with the Offset Coding Method.

In Fig. 3.18, plots (c) and (d), there are four possible binary outputs, and they are

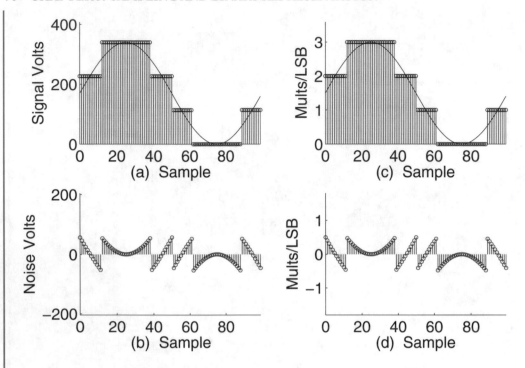

Figure 3.18: (a) A portion of an analog 10 Hz sine wave (solid), quantized using 2 bits, Offset Method, input biased one-half LSB, LSB = 113.33 volts (stem plot); (b) Quantization Noise; (c) Analog 10 Hz sine wave (solid), quantized using 2 bits, Offset Method, input biased one-half LSB, in Multiples of the LSB (stem plot); (d) Quantization Noise.

$$\begin{aligned}
\text{Binary } 0 &= [0\ 0] \quad (0 \text{ times } 113.33 \text{ Volts}) \\
\text{Binary } 1 &= [0\ 1] \quad (1 \text{ times } 113.33 \text{ Volts}) \\
\text{Binary } 2 &= [1\ 0] \quad (2 \text{ times } 113.33 \text{ Volts}) \\
\text{Binary } 3 &= [1\ 1] \quad (3 \text{ times } 113.33 \text{ Volts})
\end{aligned}$$

Note: For the following two examples, assume proper anti-aliasing filtering is used prior to any sampling and that no bias is applied to the input signal.

Example 3.19.   An analog signal $x = \cos(2\pi(100)t)$ with t in seconds is sampled for one second at 200 Hz, with the first sample being taken at time 0, when the signal value is 1.0 volts. (a). What is the value of the analog signal at the first four sampling times? (b). For the same signal, suppose the sampling rate had been 400 Hz. In that case, what would have been the values of the analog signal at the first four sampling times? (c). Assume that the sampling is done by an ADC that accepts input voltages between 0 and 1 volt. What signal adjustments need to be made prior to applying the analog signal to the ADC's input?

(a) The call

$$x = \cos(2*pi*100*(0:0.005:0.015))$$

gives the answer as [1,-1,1,-1];

(b) The call

$$x = \cos(2*pi*100*(0:0.0025:0.0075))$$

gives the answer as [1,0,-1,0];

(c) To convert the analog input signal into one which matches the ADC's input voltage range, either multiply the signal by 0.5 and then add 0.5 volt to the result, or add 1.0 volt to the signal and multiply the result by 0.5.

Example 3.20.   Assume the ADC in the preceding example, part (c), quantizes all voltages as positive offsets from 0 volts, and that a 1 volt input to the ADC yields a binary output of [1111]. Assume that the test signal is 100 Hz but that the sampling rate has been increased to 10, 000 Hz, giving rise to quantized sample values covering the entire 0 to 1 volt range. (a). If the binary output were [0001], what quantized input voltage to the ADC would be represented? (b). What range of analog signal voltages input to the ADC could have resulted in the binary output given in (a)? (c). What was the original signal voltage equivalent to the voltage determined in (a) above (remember that we had to shift and scale the original analog voltage prior to applying it to the ADC)?

(a) Since binary (offset) output [1111] (15 in decimal) is equivalent to a 1-volt analog input, it follows that binary output [0001] is equivalent to 1/15 volt;

(b) Since there is no bias to the input of 0.5 LSB, values of 1/15 volt up to just less than 2/15 volt will quantized to [0001];

(c) Since we added 1 volt and multiplied by 0.5 to get the analog input voltage, we take 1/15 volt, multiply by 2, and then subtract 1.0 volt to get -13/15 volt.

The following examples illustrate determination of the equivalent LSB voltage for a given ADC operation, and choosing the number of quantization bits to achieve a certain LSB voltage.

Example 3.21.   Suppose that we have an ADC which will accept analog input voltages ranging between −170 and +170 volts, and that is capable of quantizing the input to 3 bits, one of which will be a sign bit, leaving two bits for magnitude. Determine the value of the LSB in volts.

The maximum is 170 volts, and we will have $2^3 - 1 = 7$ distinct levels (remember that we lose a level when using one bit for the sign) which can be specified by the 3 bit output. Note that we have three distinct magnitudes above zero available for positive values, and the same for negative values, and one level for zero, giving us $3 + 3 + 1 = 7$ distinct levels of quantization.

Doing the arithmetic, we can see that one bit is equivalent to (170 volts/3 levels) = 56.67 volts/level.

Figure 3.19, plots (a) and (b), show the result. The 3-bit quantized values differ by as much as one-half the LSB (56.67 divided by two, or 28.33 volts) from the correct value. In some applications, this might be of sufficient accuracy, but for "high end" applications, such as audio and video, 16 or more bits of quantization are necessary.

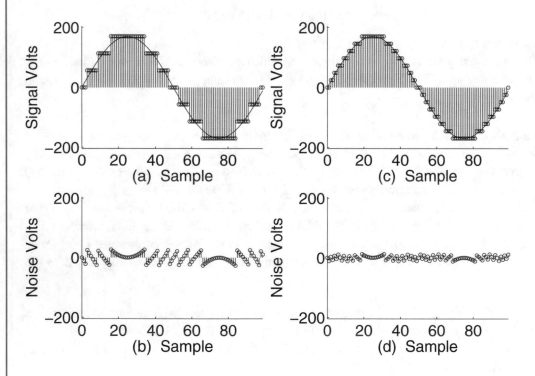

Figure 3.19: (a) A portion of an analog 10 Hz sine wave (solid), quantized using 3 bits (1 Sign, 2 Magnitude), input biased one-half LSB, LSB = 113.33 volts (stem plot); (b) Quantization Noise; (c) Analog 10 Hz sine wave (solid), quantized using 4 bits (1 Sign, 3 Magnitude), input biased one-half LSB (stem plot); (d) Quantization Noise.

Example 3.22.   Suppose that it was required that the samples be accurate to the nearest 1.25 volts. Determine the total number of bits needed to achieve this.

We would need (170 volts/1.25 volts/level) = 136 levels per side, or 273 levels total. Investigating powers of two, we see that $2^8 = 256$ is inadequate, but $2^9 = 512$ will work—so, in addition to the sign bit, we need to employ nine bits of quantization, which will give the LSB as (170 volts/255 levels) = 0.667 volts/level.

Example 3.23.   Suppose we wanted to quantize, using 3 bits and the Offset method, a 340 volt peak-to-peak sine wave (this would correspond to a sine wave varying between -170 and 170 volts which has been offset toward the positive by 170 volts such that there are no longer any negative voltages). What would be the value of the LSB? State the possible binary outputs and the corresponding input voltages they represent.

With Offset Coding, we get 2^3 distinct levels, one of which is the lowest level, zero. There are therefore $2^3 - 1 = 7$ levels to reach the highest level, which would yield the LSB as $340 \div 7 =$ 48.57 volts.

The possible binary output codes would be 000, 001, 010, 011, 100, 101, 110, 111, which would represent quantized voltages of 0, 48.57, 97.14, 145.71, 194.28, 242.85, 291.42, and 340.00 volts, respectively.

For simple input signals, the error signal is well-correlated with the input signal, as can be seen in Fig. 3.19, plots (c) and (d); when the input signal is complicated, the error signal will be less correlated with the input, i.e., appear more random. Noise that correlates well with the input tends to be tonal and more audible; there exist a number of methods to "whiten" the spectrum of the sampled signal by decorrelating the sampling noise. One such method, **Dithering**, is to add a small amount of random noise to the signal, which helps decorrelate sampling artifacts. Extensive discussions of dithering are found in [1] and [2].

Quantization using 4 bits still results in objectionably large amounts of quantization noise, as shown in Fig. 3.19, plots (c) and (d).

Note that no value of the signal to be quantized is ever further than one-half the LSB amount from one of the possible quantized output levels. This can be seen clearly in Fig. 3.18, plots (c) and (d). Recall that the LSB amount would be (170 volts/7 levels) = 24.2857 volts/level. Note that in Fig. 3.19, plot (d), the maximum noise amplitude is about half that, or a little over 12 volts.

As a good approximation, it can be said that each time the number of quantizing bits goes up by one, the maximum magnitude of quantization noise is halved, and the corresponding noise power is quartered, since power is proportional to the square of voltage or amplitude.

References [1] and [2] give extensive and accessible discussions of quantization noise.

3.13   DAC WITH VARIABLE LSB

The eventual goal of much signal processing, such as that of audio or video signals, is to convert digital samples back into an analog signal. This is the process of signal reconstruction, which is performed by sending the digital samples (binary numbers) to a DAC.

In addition, as we have seen, the heart of a successive approximation converter is a digital-to-analog converter.

Figure 3.20 depicts a typical arrangement for converting a sequence of digitized samples back into an analog signal. The latch holds a digital word (typically 4-24 bits) on the input of the DAC until the next digital word is issued from the digital sample source. Holding a value constant until the next one is retrieved is referred to as a **Zero Order Hold**, which is discussed later in this chapter.

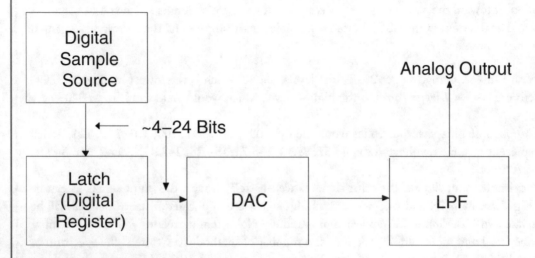

Figure 3.20: A representative arrangement for converting digitized samples to an analog signal. The digital path to the DAC is usually a number of bits as depicted, presented in parallel to the DAC, which outputs an analog signal on a single line.

A simple DAC is shown Fig. 3.21. A reference voltage feeds binarily-weighted resistors that feed the inverting input of an operational amplifier, which sums the currents and outputs a voltage proportional to the current according to the formula

$$V_{out} = V \cdot (Bit(3)(-R/R) + Bit(2)(-R/2R) + Bit(1)(-R/4R) + Bit(0)(-R/8R))$$

which reduces to

$$V_{out} = -V \cdot (Bit(3) + 0.5Bit(2) + 0.25Bit(1) + 0.125Bit(0))$$

where $Bit(0)$ means the value (1 or 0) of the 2^0 bit, and so forth. This scheme can be expanded to any number of needed bits.

Here we see that for the type of DAC shown in Fig. 3.21,

$$LSB_{volts} = -V/(2^{N-1}) \tag{3.2}$$

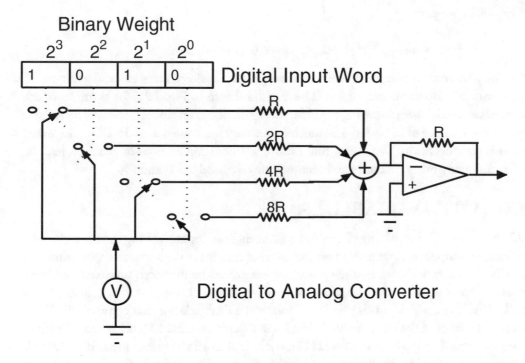

Figure 3.21: A simple 4-bit Digital-to-Analog converter. Note that the contributions from each resistor enter a summing junction, the output of which is connected to the negative input of the operational amplifier; the summing junction is shown for conceptual clarity only. In an actual device, the summation would be accomplished simply by connecting the leads from all resistances directly to the negative input of the operational amplifier.

where N is the number of quantization bits; hence the LSB value may be arbitrarily set by changing the reference voltage V. We can also note that when all bits are equal to 1, the maximum value of V_{out} is attained, and

$$LSB_{volts} = -(V_{out})_{max}/(2^N - 1) \tag{3.3}$$

Example 3.24.   A sequence that is in four-bit offset binary format is to be converted to the analog domain using a DAC of the type shown in Fig. 3.21, and it is desired that its analog output voltage range be 0 to +13.5 volts. Determine the necessary value of reference voltage V.

We note that for four bits, the highest level is 15 times the LSB value, so we get

$$LSB_{volts} = 13.5/15 = 0.9 \text{ volt}$$

and using Eq. (3.2), we get

$$V = -(2^{N-1})LSB_{volts} = -8(0.9) = -7.2 \text{ volt}$$

There are a number of alternative methods for digital-to-analog conversion, and a number of other refinements and aspects to conversion. The method shown in Fig. 3.21, for example, which uses resistors having resistances following the binary weights, places severe requirements on resistor accuracy when the number of bits is large. A much better method, known as an R-2R ladder, exists which uses only resistors having two values, thus making accurate implementation much easier. The R-2R ladder, and many other structures and methods are discussed in [2] and [3].

3.14   ADC WITH VARIABLE LSB

Figure 3.22 shows an ADC consisting of a comparator having one input receiving the analog sample to be quantized, an output which is used to set and/or reset each bit in the successive approximation register in accordance with the successive approximation procedure discussed previously, a binary weight generator the values of which are equal to (reading from left to right) 8, 4, 2, and 1 times a reference LSB voltage, which in this case is 1.0 volt, and a feedback line that connects the DAC output to a second input of the comparator. At the instant shown, sample 12 (analog value 15 volts) has just been quantized to the binary value [1111], equivalent to 15 volts since the LSB is 1.0 volt (for this example, no offset bias has been applied to the input, so all quantized values are at or below the sample values).

In Fig. 3.23, the LSB voltage has been changed to 0.75 volt. Since four bits are being used, the maximum voltage output of the DAC section is 15 times 0.75 = 11.25 volt. All input signal values at or above 11.25 volt will therefore be quantized at the highest binary output of the ADC, which is [1111]. Note that the Binary Weight Generator now bears weights proportional to the LSB value of 0.75 volt, i.e., 6, 3, 1.5, and 0.75 volt.

Figures 3.22 and 3.23 were created using the script

ML_SuccessiveApproximation

which, when called, opens up the GUI as shown in the figures just mentioned. The GUI allows selection of a number of parameters including automatic (fast or slow) or manual computation.

3.15   ZERO-ORDER HOLD CONVERSION

Figure 3.24 shows what the output of a DAC with a zero-order hold would look like prior to filtering with a low pass filter. In the Zero-Order Hold system, each DAC conversion value is held on the output of the DAC until the next sample arrives at the DAC and changes the output value.

The use of a zero-order hold in digital-to-analog conversion is perhaps the simplest or most practical method of conversion. The stairstep waveforms it generates contain the original sinusoids,

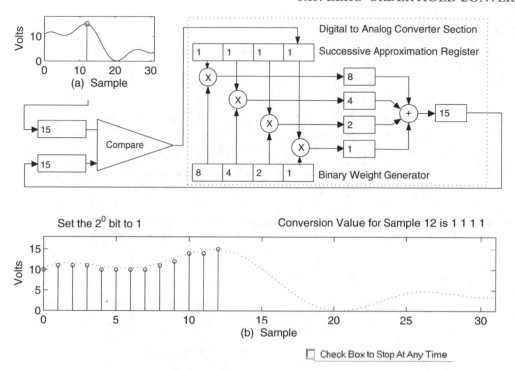

Figure 3.22: A schematic arrangement of an ADC having LSB = 1.0 volt, using the Offset method with no bias to the input signal; (a) Analog signal (solid), sample being quantized (circle); (b) Quantized samples.

plus overtones thereof (for a basic discussion of conversion methods other than by use of a zero-order hold, see reference [1]).

There are actually two entirely separately arising spectral impurities (frequency components that were not present in the original analog signal) in the output of a DAC having a zero-order hold: quantization noise, and stairstep noise. Even if the number of bits of quantization is very large, leading to effectively no quantization error, there will still be a very large stairstep component in the DAC output which was not in the original analog signal. Referring to a sine wave, for example, the amplitude of the stairstep component is large when the number of samples per cycle of the sine wave is low, and decreases as the number of samples per cycle increases.

Fortunately, the frequencies comprising the stairstep component all lie above the Nyquist limit and can be filtered out using a lowpass reconstruction filter designed especially for zero-order hold reconstruction (LPF in Fig. 3.20). Quantization noise, on the other hand, has components below the Nyquist limit, and thus cannot be eliminated with lowpass filtering. Zero-order hold conversion acts like a gently-sloped lowpass filter, so a special reconstruction filter must be used that cuts off at the Nyquist limit but emphasizes higher frequencies in the passband. Details may be found in [2].

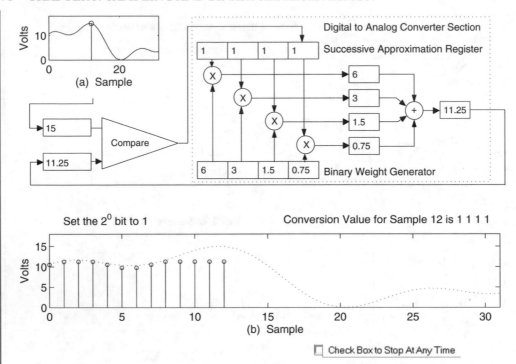

Figure 3.23: A schematic arrangement of an ADC having LSB = 0.75 volt, using the Offset method with no bias to the input signal; (a) Analog signal (solid), sample being quantized (circle); (b) Quantized samples.

3.16 CHANGING SAMPLE RATE

There are many different standard sample rates in use, and from time to time it is necessary to convert one sequence sampled at a first rate to an equivalent sequence sampled at a different rate. It might, for example, be necessary to add two sequences together, one of which was sampled at 11.025 kHz, and the other sampled at 22.05 kHz. To do this, one of the sequences must have its sample rate converted to match that of the other prior to addition.

In this section we look at two important processes, interpolation and decimation by whole number ratios, and the combination of the two techniques, which allows change of sample rate by a whole number ratio, such as 4:3, for example. We include a basic discussion of interpolation using the *sinc* function, and compare it to linear interpolation.

3.16.1 INTERPOLATION

From time to time, it is necessary to either increase the sampling rate of a given sequence, or decrease the frequencies contained in the sequence by a common factor. Let's consider a concrete example.

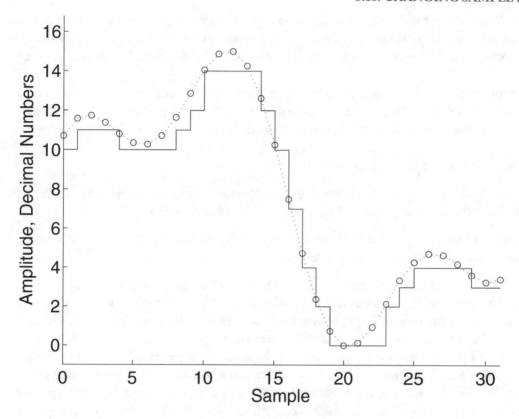

Figure 3.24: Original Analog Signal (Dashed); Ideal (i.e., Quantized to Infinite Precision) Samples (Circles); Quantized Samples at DAC output (Solid).

Supposing one second's worth of samples were taken at a sampling rate of 65,536 Hz of a signal composed of three sine waves having respective frequencies of 3000 Hz, 4000 Hz, and 5000 Hz (any other collection or mixture of frequencies would have been possible provided, of course, that the highest frequency was less than half of 65,536).

Now suppose we would like to divide all of these frequencies by a common factor, such as 10. This would reduce the three frequencies to 300, 400, and 500 Hz.

Thinking intuitively, one cycle of a sine wave occurring over 32 samples (for example), can have its apparent frequency reduced by a factor of 10 (for example) if something were done to make the same one cycle occur over 320 samples instead of 32, assuming that the sample rate remains the same. Thus it would now take 10 times as long to read out that one sine wave cycle, hence its frequency is now one-tenth of the original frequency.

If, on the other hand, we only want to increase the number of samples per cycle while keeping the same frequency, we must increase the sample rate by the same factor with which we increased the number of samples in the sequence. For example, in order to make the sine wave with 320 samples per cycle sound at the same frequency as the original 32-sample sine wave when played out through an audio system, it will be necessary to send samples to the DAC at 10 times the original sampling rate.

In many cases, what is needed is to convert a signal having a first sample rate to a second sample rate, where the frequencies upon final readout (or usage) are the same. In that kind of situation, if the final sample rate were higher, then it would obviously be necessary to have more samples per cycle.

Here are two simple rules to remember:

• To double the sample rate while keeping the frequencies the same, double the number of samples in the sequence (we'll see how shortly) and double the sample rate.

• To halve the frequency content (divide all frequencies by two), double the number of samples in the sequence and keep the sample rate the same.

Now we'll look at how to actually increase the number of samples in the sequence.

An easy and straightforward way to accomplish this operation, called interpolation, is to insert extra samples, each with a value of zero, between each of the original sequence's samples, and then lowpass filter the resultant sequence. This is called **Zero-Stuffing**.

The cutoff frequency of the lowpass filter should be set at the reciprocal of the interpolation factor. Recall that for a digital filter, all frequencies are normalized, with 1.0 representing the Nyquist limit frequency. Thus, if we were interpolating by a factor of four, for example, the lowpass filter cutoff would be set at $1/4 = 0.25$. This choice of cutoff frequency passes only the original frequency content, as effectively decreased by the interpolation factor.

The script

$$LV\_InterpGeneric(Freq, InterpFac)$$

demonstrates the principles involved.

Example 3.25.   Demonstrate interpolation by a factor of 2 on a 4-cycle sine wave over 32 samples.

We call the script as follows:

LV_InterpGeneric(4,2)

This will show four cycles of a sine wave, as shown at (a) in Fig. 3.25, and will double the number of samples per cycle using the zero-stuffing technique. Plot (b) of Fig. 3.25 shows the original sequence with zero-valued samples inserted between each pair of original samples, and Plot (c) of Fig. 3.25 shows the result after lowpass filtering.

Figure 3.26 shows what happens from the frequency-domain point of view: at (a), where we see that the spectrum shows only a single frequency (4 Hz), corresponding to the four cycle sine

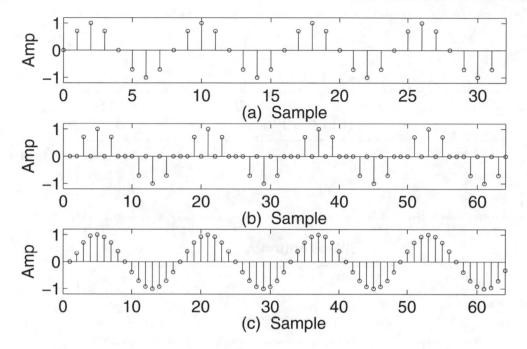

Figure 3.25: (a) Original sequence; (b) Sequence at (a) with one zero-valued sample inserted between each original sample; (c) Sequence at (b) after being lowpass-filtered.

wave shown in plot (a) of Fig. 3.25. After zero-stuffing, another frequency component appears at 28 Hz (plot (b) of Fig. 3.26). Thinking logically, we want only the 4 Hz component in the final sequence, so lowpass filtering to get rid of the 28 Hz should restore the signal to something that looks like a pure sine wave, which in fact it does. Plot (c) of Fig. 3.26 shows the resultant spectrum, which shows a few small amplitude components at frequencies other than four. These components correspond to the distortion which may be seen on the leading edge of the signal shown in plot (c) of Fig. 3.25. This distortion is merely the transient, nonvalid output of the lowpass filter used to eliminate the high frequency (28 Hz in this example) component.

Example 3.26.   Demonstrate interpolation of a multi-frequency audio signal. Play the interpolated sequence at the original sample rate to show the frequency decrease, and play it at the new sample rate necessary to maintain the original output frequencies.

The script

LV_InterpAudio(InterpFac)

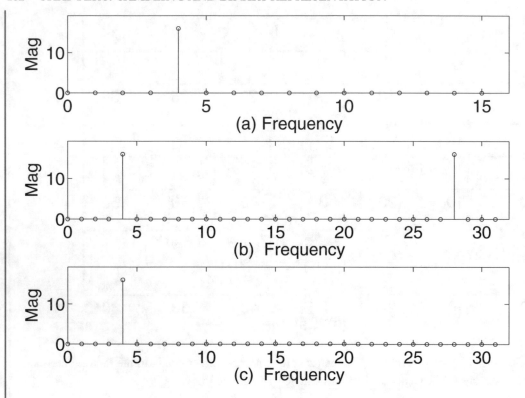

Figure 3.26: (a) Spectra of original sequence as shown in the previous figure; (b) Spectrum of the zero-stuffed sequence as shown in (b) in the previous figure; (c) Spectrum of the lowpass-filtered sequence as shown in (c) of the previous figure.

allows you to interpolate an audio signal which is a mixture of frequencies 400, 800, and 1200 Hz, sampled at 11025 Hz, and play the results. The value of input argument *InterpFac* is limited to 2 and 4.

A typical call might be

LV_InterpAudio(2)

which will stuff one sample valued at zero between each sample of the original test sequence, and then lowpass filter using a normalized cutoff frequency of 1/2.

Figure 3.27, in plot (a), shows the first 225 samples of the original sequence, about 8 cycles of the repetitive, composite signal, of which the lowest frequency is 400 Hz. Plot (b) of Fig. 3.27 shows the first 225 samples of the interpolated sequence, showing about four cycles of the frequency-reduced composite signal (now with the lowest frequency being 200 Hz).

Plot (a) of Fig. 3.28 shows the frequencies in the original sequence as 400, 800, and 1200 Hz; in plot (b), we see that the three frequencies have indeed been reduced by a factor of 2 to 200, 400,

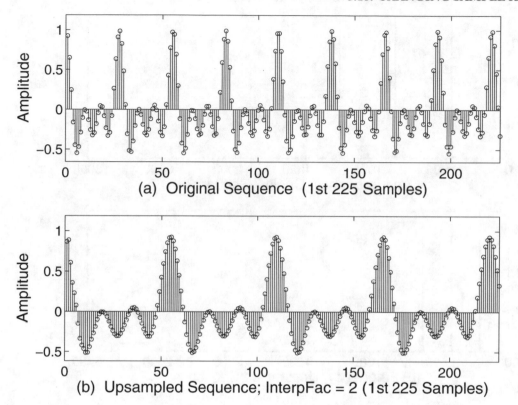

Figure 3.27: (a) A portion of the original sequence; (b) The same number of samples of the interpolated sequence.

and 600 Hz. In general, the frequencies in the spectrum of the upsampled sequence are decreased by the interpolation factor, assuming that the sampling rate remains the same. If the sampling rate were also increased by the same factor as the interpolation factor, then the frequencies would not change; there would merely be a larger number of samples per cycle.

For MATLAB users only, the script

$$ML\_InterpAudioPOC(InterpFac)$$

is very similar to $LV\_InterpAudio$, except that $InterpFac$ may assume the positive integral values 2, 3, 4, etc. Several of the windows that are opened by $ML\_InterpAudioPOC$ have push buttons that allow playing the input sequence at the original sample rate (11025 Hz) and the output sequence at the original or new sample rates.

- To maintain the same apparent output frequency after interpolation, compute the necessary post-interpolation sample rate SR_{New} as follows:

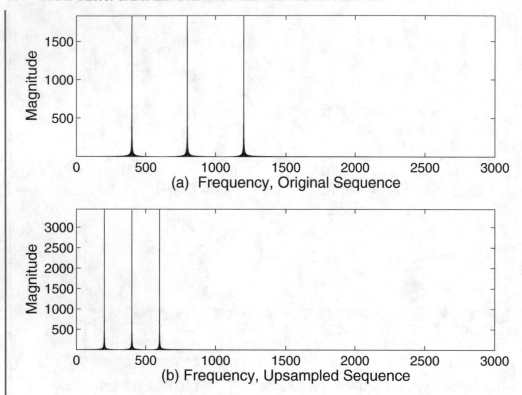

Figure 3.28: (a) Spectrum of the original signal; (b) Spectrum of the interpolated signal, showing a frequency reduction of a factor of two.

$$SR_{New} = SR_{Orig} \cdot I$$

where I is the interpolation factor and the original, pre-interpolation sample rate is SR_{Orig}.

- If the sample rate is not to be changed, then the post-interpolation apparent output frequency F_{New} is computed as follows:

$$F_{New} = \frac{F_{Orig}}{I}$$

where the original frequency is F_{Orig}.

3.16.2 DECIMATION

For the problem of upsampling or interpolation, we inserted extra samples between existing samples, and then lowpass filtered to limit the frequency content to that of the original signal.

For the problem of downsampling or decimation, a new sequence is created consisting of every n-th sample of the original sequence. This procedure decreases the sampling rate of the original signal, and hence lowpass filtering of the original signal must be performed prior to decimating.

Example 3.27. We have 10,000 samples, obtained at a sample rate of 10 kHz. Suppose the highest frequency contained in the signal is 4 kHz, and we wish to downsample by a factor of four. Describe the necessary filtering prior to decimation.

The new sampling rate is 2.5 khz. Thus the frequency content of the original signal must be filtered until the highest remaining frequency does not exceed one-half of 2.5 kHz, or 1.25 kHz. Thus the original signal must be severely lowpass filtered to avoid aliasing.

Example 3.28. Consider a different example, in which the original sequence is 50,000 samples, sampled at 50 kHz, with the highest frequency contained therein being 4 kHz. We wish to decimate by a factor of four. Describe the necessary filtering prior to decimation.

The new sample rate is 12.5 kHz, which is more than twice the highest frequency in the original sequence, so in this case we do not actually have to do any lowpass filtering prior to decimation, although lowpass filtering with the proper cutoff frequency should be conducted as a matter of course in any algorithm designed to cover a variety of situations, some of which might involve the new Nyquist rate falling below the highest signal frequency.

The following script allows you to decimate a test signal having multifrequency content.

$$LV\_DecimateAudio(DecimateFac)$$

The argument *DecimateFac* is the factor by which to decimate, and it may have values of 2 or 4. This is based on the use of a test signal comprising frequencies of 200, 400, and 600, and an original sampling rate of 44,100. The function

$$sound(y, Fs)$$

(where y is a vector of audio samples to be played, and Fs is the sample rate) is used to play the resultant sounds, and, in LabVIEW, it permits only a few sample rates, namely: 44,100 Hz, 22,050 Hz, 11,025 Hz, and 8,000 Hz. The script creates the original signal at a sample rate of 44,100 Hz, and allows you to decimate by factors of 2 or 4, thus reducing the sample rates to 22,050 Hz and 11,025 Hz, respectively. The script automatically sounds the original signal at 44,100 Hz, the decimated signal at 44,100, and then the decimated signal at the new sample rate, which is either 22,050 Hz or 11,025 Hz depending on whether *DecimateFac* was chosen as 2 or 4.

Figure 3.29 shows a result from the script call

LV_DecimateAudio(2)

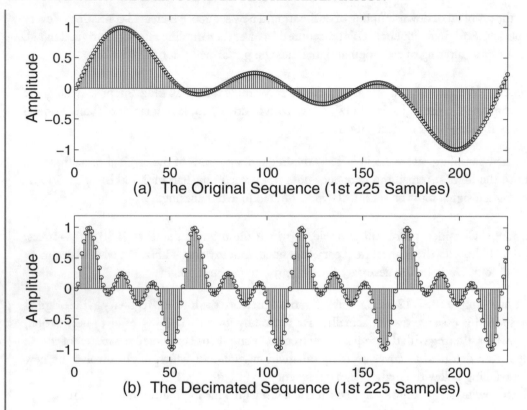

Figure 3.29: (a) Original sequence; (b) Result from taking every 2nd sample of sequence at (a), after suitably lowpass filtering.

which decimates the test audio signal by a factor of 2.

Decimation of a complex waveform can be used to decrease the number of samples per cycle, or increase the output frequency, depending on whether the sample rate decreases proportionately to the decimation factor, or whether it remains at the original rate. In the former case, the apparent output frequency is the same, but with fewer samples per cycle; in the latter case, the apparent output frequency increases by the decimation factor.

For MATLAB users, the script

$$ML\_DecimateAudioPOC(DecimateFac)$$

is similar to $LV\_DecimateAudio$, except that *DecimateFac* may take on the positive integral values 2, 3, 4, etc. Several of the windows opened by the scripts provide push buttons that may be used to play on command the original and decimated sequences at the original sample rate.

- To maintain the same apparent output frequency after decimation, compute the necessary post-decimation sample rate SR_{New} as follows:

$$SR_{New} = \frac{SR_{Orig}}{D}$$

where D is the decimation factor and the original, pre-decimation sample rate is SR_{Orig}.

- If the sample rate is not to be changed, then the post-decimation apparent output frequency F_{New} is computed as follows:

$$F_{New} = F_{Orig} \cdot D$$

3.16.3 COMBINING INTERPOLATION WITH DECIMATION

Based on what we've learned about interpolation and decimation, it should be obvious that we can alter sample rate by (theoretically) any whole number ratio. For example, if the original sample rate is 44,100 Hz, and we happened to want to change it to 33,075 Hz, this could be done by interplating by a factor of 3, and then decimating by a factor of 4. The steps must be done in this order: interpolation first, lowpass filtering, then decimation. In this way, only one lowpass filtering operation is necessary, after interpolation but before decimation. The lowpass filter cutoff is chosen as the more restrictive of the cutoff frequencies of the two lowpass filters that would have been required by the interpolation and decimation operations had they been performed by themselves.

The script

$$LV\_ChangeSampRateByRatio(OrigSR, InterpFac, DecimateFac)$$

allows for experimentation in changing sample rate using this method. If the value of *OrigSR* is one of the standard sampling rates, namely, 8 kHz, 11.025 kHz, 22.05 kHz, or 44.1 kHz, the original signal will be played as an audio signal. Likewise, if the final sample rate, computed according to the input arguments, is within 1 % of one of the standard sampling rates, the final or resampled signal will also be played out as an audio signal using the function *sound* (see the script for details). A typical call would be

LV_ChangeSampRateByRatio(8000,22,8)

which interpolates by a factor of 22, lowpass filters (with a normalized cutoff frequency equal to the reciprocal of the greater of the interpolation factor and the decimation factor, which in this case is 22), and then decimates by a factor of eight. Figure 3.30 shows leading portions of the input and output waveforms. The input waveform is a test signal consisting of three sinusoids have frequencies of 400, 800, and 1200 over a sequence length of 8000.

For MATLAB users, the script

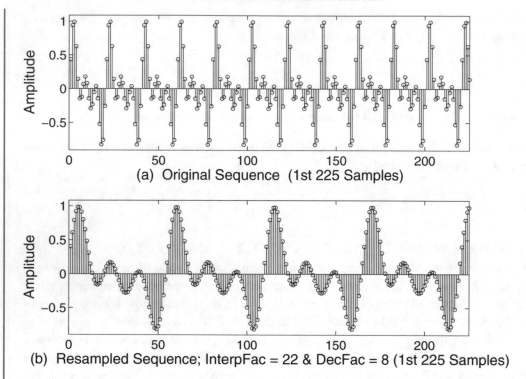

Figure 3.30: (a) Original sequence; (b) Result from interpolating by a factor of 22 and decimating by a factor of eight.

$$ML\_ChangeSampRateByRatioPOC(InterpFac, DecimateFac)$$

is similar to $LV\_ChangeSampRateByRatio$, except that the original sample rate is set at 44,100 Hz, and the final sample rate, which is (InterpFac/DecimateFac), may be other than a standard sampling rate (as mentioned above).

MathScript has sophisticated functions for interpolation, decimation, and the combination of the two, called resampling. The functions are *interp*, *decimate*, and *resample*. These functions take great pains to give the most accurate result possible, especially at the first and last samples of an input sequence. In using any of these functions, for example, you'll find that the output sequence is temporally aligned as best as possible with the input; that is to say, the waveform phase of the output sequence matches the waveform phase of the input sequence as closely as possible. Also, the lowpass filter length is made variable according to the input parameters.

3.16.4   BANDLIMITED INTERPOLATION USING THE SINC FUNCTION

According to the Shannon sampling theorem, a bandlimited continuous domain signal can be exactly reconstructed from its samples provided the sampling rate is at least twice the highest frequency contained in the bandlimited signal. Then

$$x(t) = \sum_{n=-\infty}^{\infty} x(nT_S) \frac{\sin((\pi/T_S)(t - nT_S))}{(\pi/T_S)(t - nT_S)}$$

Here the sinc functions are continuous functions ranging over all time, and exact reconstruction depends on all the samples, not just a few near a given sample. In practical terms, however, it is possible to reconstruct a densely-sampled version of the underlying bandlimited continuous domain signal using discrete, densely-sampled windowed sinc functions that are weighted by a finite number of signal samples and added in an offset manner. Such a reconstruction provides many interpolated samples between the signal samples, and thus is useful in problems involving change of sample rate.

The script (see exercises below)

$$LVxInterpolationViaSinc(N, SampDecRate, valTestSig)$$

creates a densely sampled waveform (a simulation of a continuous-domain signal) of length N, then decimates it by the factor *SampDecRate* to create a sample sequence from which the original densely sampled waveform is to be regenerated using sinc interpolation. Two methods are demonstrated.

One method is to create a sinc function of unit amplitude having a period of M samples and an overall length of perhaps $5*M$ or more samples. The sinc function is then windowed. A sample sequence to be used to reconstruct a densely sampled version of its underlying bandlimited continuous domain signal has one-half the sinc period, less one, zeros inserted between each sample, and the zero-padded sequence is then linearly convolved with the sinc function. This is essentially the zero-stuffing method using a sinc filter as the lowpass filter. This method is no more efficient than linear convolution. It computes many values between samples, when, in most applications, only one or a few interpolated values are needed between any two samples.

Another method, which allows you to obtain a single interpolated value between two samples of a sequence is matrix convolution. A matrix is formed by placing a number of copies of a densely-sampled windowed sinc function as columns in a matrix, each sinc function offset from the previous one by $M/2$ samples (it is necessary to let M be even). A particular row of the matrix is selected (according to exactly where between two existing samples another sample value is needed) and then multiplied by the sample sequence (as a column vector) to give an interpolated value.

Figure 3.31, which results from the call

LVxInterpolationViaSinc(1000,100,1)

illustrates the matrix convolution method. For this example, which uses 10 contiguous samples from a sequence to perform a bandlimited, densely-sampled reconstruction of the underlying continuous domain signal, there are 10 matrix columns containing sinc functions, offset from each other.

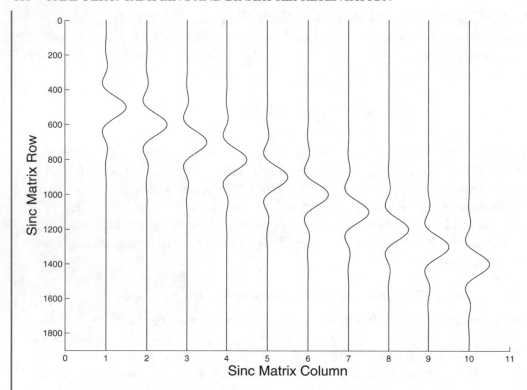

Figure 3.31: Sinc functions, offset from one another by 100 samples, depicted in 10 columns of a matrix suitable for bandlimited interpolation between samples of a sequence. Note that for display purposes only, all sinc amplitudes have been scaled by 0.5 to avoid overlap with adjacent plots. They are not so scaled in the actual matrix in the script.

Figure 3.32 shows the matrix columns after being weighted by the appropriate samples, and at column zero, the sum of all the weighted columns is shown, which forms the reconstruction.

Figure 3.33 shows more detail of the reconstruction: the simulated continuous-domain signal (solid line), 10 samples extracted therefrom by decimating the original signal by a factor of 100 (stem plot), and the reconstruction (dashed line) of the original signal according to the matrix convolution method described above.

The results of the linear convolution method are shown in Fig. 3.34. Note again that the best approximation to the original signal is found in the middle of the reconstruction.

Reference [4] discusses ideal sinc interpolation in Section 4.2, and covers sampling rate conversion in Chapter 10.

Example 3.29.   Devise a simple (not necessarily efficient) method to convert a sequence sampled at 44,100 Hz to an equivalent one sampled at 48,000 Hz.

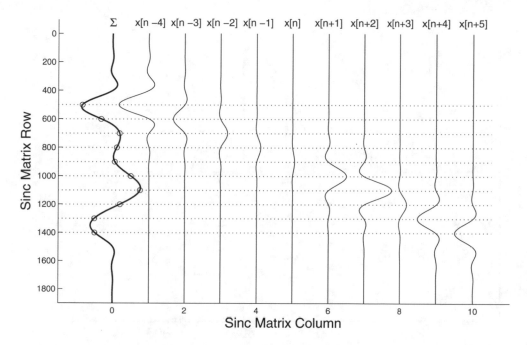

Figure 3.32: Sinc matrix columns, each weighted by the sample shown at the top of the column, and the sum of all columns (i.e., the interpolated reconstruction), at the left with the symbol Σ at the column head. The sample values corresponding to samples x[n-4] to x[n+5] are plotted as circles on the reconstruction, with x[n-4] being topmost. The dashed horizontal lines intersect the 10 matrix columns at the values that are summed to equal the corresponding interpolated value marked with a circle; the particular rows summed correspond to the sample values x[n-4] up to x[n+5]. Any row may be summed (as a single operation) to obtain an interpolated value between the original samples. Note that the smallest reconstruction errors occur in the middle of the reconstruction. Note also that all signal amplitudes have been scaled by 0.5 to avoid overlap with adjacent plots, but are not so scaled in the computation.

The sample rate ratio can be approximated as 160:147. Assuming that a sequence *Sig* sampled at 44.1 kHz has indices 1,2,3, ..., the indices of samples, had the sample rate been 48 kHz instead of 44.1 kHz, would be

$$1:147/160:length(Sig) = 1, 1.9188, 2.8375, 3.7562...$$

This generates a total number of new samples of 160/147*length(Sig), which, when read out at the new sample rate of 48 kHz, will produce the same apparent frequencies as the original sequence read out at 44.1 kHz. To obtain the sample values, sinc interpolation as discussed above can be used. The more densely sampled the sinc function, and the larger the number of original sequence samples included in the approximation, the more accurate will be the interpolated values.

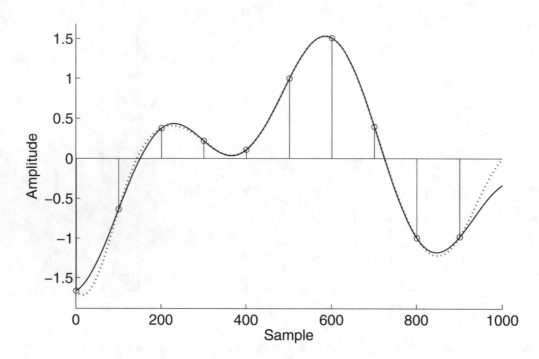

Figure 3.33: Continuous-domain signal (solid), simulated with 1000 samples; 10 samples of the (simulated) continuous-domain signal (stem plot); a reconstruction (dashed line) of the continuous-domain signal from the 10 samples. Note that the reconstruction is only a good approximation from about samples 400-700 of the original (simulated) continuous-domain signal.

The script (see exercises below)

$$LVxInterp8Kto11025(testSigType, tsFreq)$$

performs both linear and sinc interpolation of an audio signal to change its sample rate from 8 kHz to 11.025 kHz. The audio signal may be a cosine having a specified frequency, or the file 'drwatsonSR8K.wav'.

Figures 3.35 and 3.36, respectively, show several samples from the interpolation of cosines of 250 Hz and 2450 Hz, respectively, obtained by making the calls

LVxInterp8Kto11025(0,250)
LVxInterp8Kto11025(0,2450)

Note that at the much lower frequency (i.e., low relative to the Nyquist rate), the sinc- and linear-interpolated samples do not differ much, whereas at the higher frequency, there are large discrepancies.

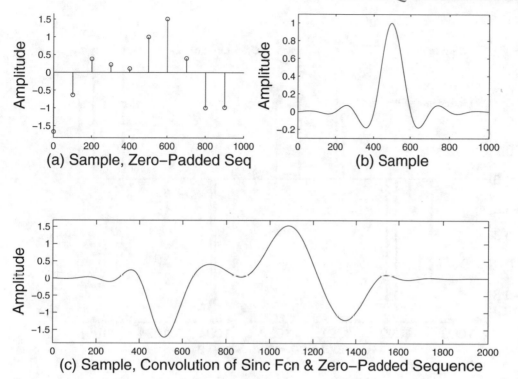

Figure 3.34: (a) The sample sequence, padded with 99 zeros between each sample; (b) The *sinc* function; (c) Convolution of waveforms at (a) and (b).

3.16.5 EFFICIENT METHODS FOR CHANGING SAMPLE RATE

More advanced (i.e., efficient) methods of interpolation exist which make interpolation and decimation by ratios of large numbers (such as 160:147) feasible for real time processing. References [7]-[9] discuss efficient sample rate conversion methods.

3.17 FREQUENCY GENERATION

Sinusoids of arbitrary frequency for use in music or the like can be generated by storing one or more cycles of a sinusoid in a memory and cyclically reading out samples. The effective frequency can be varied using decimation by, in general, nonintegral decimation factors. Linear interpolation between samples is often adequate, so the more costly sinc interpolation referred to above need not be used.

3.17.1 VARIABLE SR

A first method of readout is to sequentially and cyclically read out every sample in the ROM at a frequency (sample rate) which is a multiple of the number of samples in the ROM. This method

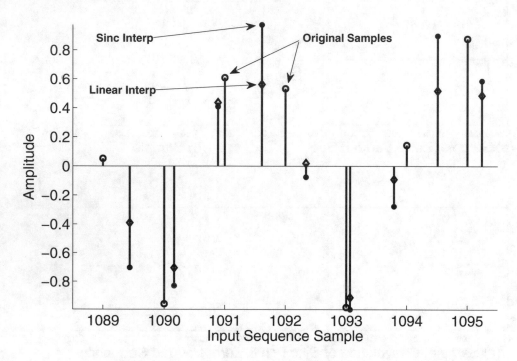

Figure 3.35: A plot of several samples from an audio sequence sampled at 8 kHz, with interpolated values at the new sample rate of 11025 Hz. Interpolated values computed using linear interpolation are plotted as stems with diamond heads, and those computed using sinc interpolation are plotted as stems with stars. The original samples, plotted at integral-valued index values, are plotted as stems headed by circles. Note that in some cases the sinc-interpolated value differs remarkably from the linear-interpolated value. The test signal consisted of a 2450 Hz cosine wave. The interpolated samples are shown prior to post-interpolation lowpass filtering.

is used principally in electronic musical instruments, where hardware or software is available to generate all the different sample rates needed. In this method, each sample of the ROM is read out in its turn, one after another. For example, if the ROM contains one cycle of a sine over 32 samples, and we want a 1 kHz output, we must read the ROM at 32 kHz. This is very straightforward. If we happened to have 2 cycles of a sinusoid in the ROM, over 32 samples, we would only need to have a sample readout rate of 16 kHz.

3.17.2   CONSTANT SR, INTEGRAL DECIMATION

Another method of readout from a ROM containing a sinusoid is to maintain a fixed sample rate SR, and to read out every n-th sample.

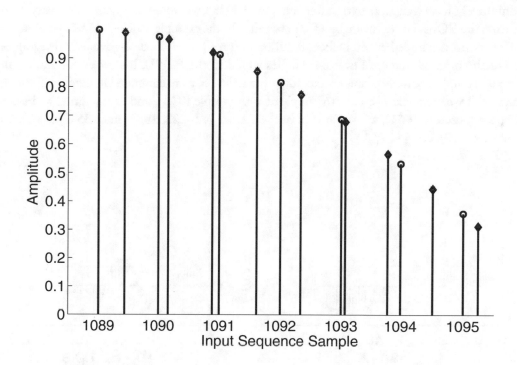

Figure 3.36: Original samples (at 8 kHz sample rate) and interpolated values (at 11025 Hz) of a 250 Hz cosine, same marker system as for the previous figure. Note that at this lower frequency there are many more samples per cycle of the signal, and the difference between linear and sinc interpolation is much less. The interpolated samples are shown prior to post-interpolation lowpass filtering.

A formula expressing the net output frequency of a sinusoid produced by reading each address sequentially from a ROM containing a sinusoid would be

$$F_{Out} = \frac{SR}{(N \, / \, F_{ROM})} = \frac{SR \cdot F_{ROM}}{N}$$

where SR is the rate at which samples are read from the ROM, N is the total number of samples in the ROM and F_{ROM} is the number of cycles of the sinusoid contained in the N samples of the ROM. Most usually, $F_{ROM} = 1$, but that need not be the case.

Consider the case of a 32-sample ROM containing one cycle of a sine wave. The ROM is sequentially read out, one address at a time, at a rate of 48 kHz, yielding a net output frequency of

$$F_{Out} = \frac{48000 \cdot 1}{32} = 1500 \text{ Hz}$$

Intuitively, if we wanted an output frequency of 4500 Hz, we would simply read out every third sample from the ROM, i.e., decimate by a factor of three. In the example shown in Fig. 3.37, plots (a) and (b), every third sample has been taken, resulting in three cycles of a sine wave over 64 samples (instead of the original one cycle over 64 samples stored in the ROM), the cycles containing an average 64/3 samples. Sample addresses exceeding the ROM size are computed in modulo-fashion. For example, if we were taking every 7th sample of a 32-sample ROM, and enumerating addresses as 1-32 (as opposed to 0-31), we would read out addresses 7, 14, 21, 28, 3 (since 35 - 32 = 3), 10 (since 42 - 32 = 10), etc.

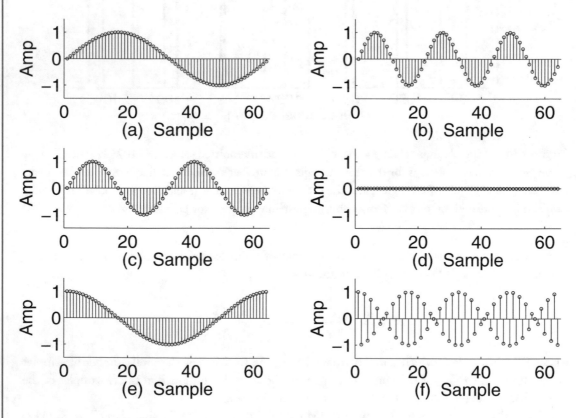

Figure 3.37: (a) A 64-sample sine wave; (b) The resultant waveform from reading out every third sample, i.e., sample indices 1, 4, 7, etc; (c) Original 64-sample sequence; (d) Decimated by a factor of 32 (only phases 0 and 180 degrees of the sine are read); (e) Original 64-sample sequence; (f) Result from taking every 34th sample.

From the above, we can therefore state a formula for the net output frequency when applying a decimation factor D.

$$F_{Out} = \frac{SR \cdot D \cdot F_{ROM}}{N} \tag{3.4}$$

The script

LVSineROReadout(N,F_ROM,D)

applies Eq. (3.4) to illustrate the principle of reading a sinusoid from a ROM using decimation. The call

LVSineROMReadout(32,1,3)

for example, results in Fig. 3.38.

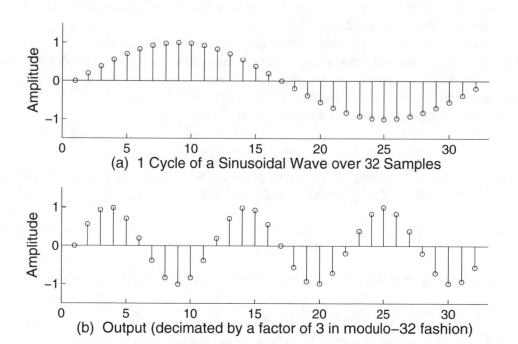

(a) 1 Cycle of a Sinusoidal Wave over 32 Samples

(b) Output (decimated by a factor of 3 in modulo–32 fashion)

Figure 3.38: (a) One cycle sine wave over 32 samples; (b) Waveform at (a), decimated by a factor of 3.

As we saw in an earlier chapter, when sampling a sinusoid at exactly its Nyquist limit (2 samples per cycle), there can be trouble when the phase is not known or controlled. You'll be well-reminded of the dangers of sampling exactly at the Nyquist limit by looking at Fig. 3.37, plots (c) and (d), which depict the result from taking every 32nd sample from two cycles of a sine wave.

Now let's decimate such that the final sample rate is less than twice the highest frequency in the ROM. If we were to decimate by a factor of 34, the number of samples per cycle would be reduced to $64/34 = 1.882$ which of course implies that aliasing will occur. The result from decimating a one cycle, 64-sample cosine by a factor of 34 is shown in Fig. 3.37, plots (e) and (f).

3.17.3   CONSTANT SR, NON-INTEGRAL DECIMATION

By removing the restriction that D be an integer in Eq. (3.4), it is possible to achieve any output frequency within the allowed range of output frequencies.

For example, let D equal a noninteger such as 2.5. Then we would have

$$F_{Out} = \frac{32000 \cdot 2.5 \cdot 1}{32} = 2500 \text{ Hz}$$

The problem with nonintegral decimation is that there are no ROM addresses other than the integers from 1 to N, the length of the ROM. For a decimation factor of 2.5, for example, we would have to read out ROM Address 1 (which exists), Address 3.5 (which doesn't exist), Address 6 (which exists), Address 8.5 (which doesn't exist), etc.

We can, however, simply compute a value for the sinusoid had the ROM Address existed, using an interpolation method, such as simple linear interpolation, the accuracy of which increases as the number of samples per cycle in the ROM increases, or a more sophisticated interpolation scheme using three or more contiguous existing values in the ROM.

For linear interpolation, the following formula applies:

$$y(x_1 + \Delta x) = y(x_1) + (\Delta x) \cdot (y(x_2) - y(x_1)) \tag{3.5}$$

where x_1 and x_2 are adjacent integral ROM addresses, and Δx lies between 0 and 1, and y is the ROM's actual or imputed value at integral or nonintegral addresses, respectively.

Example 3.30.   Set up Eq. (3.5) to determine the value of a 32-sample sine wave at nonintegral sample number 3.2 (samples being numbered 1 to 32).

Using Eq. (3.5), the value we would use for ROM Address 3.2, $y(3.2)$, would be

$$y(3.2) = y(3) + 0.2 \cdot (y(4) - y(3))$$

To explore this using Command Line calls, make the following call:

y = sin(2*pi*(0:1:31) / 32); y3pt2 = y(3) + (0.2)*(y(4) - y(3))

The above call yields *y3pt2* = 0.4173. To see what the sine value would be using MathScript, make this call:

alty3pt2 = sin(2*pi*2.2 / 32)

which yields *alty3pt2* = 0.4187 (note that the third sample of *y* is for *t* = 2 / 32, since the time vector starts at 0/32, not 1/32).

For a closer approximation using linear interpolation for a sine, start with a longer sine. For example, computing sample 3.2 for a length 512 sine make the following calls:

$$y = sin(2*pi*(0{:}1{:}511) / 512); \ y3pt2 = y(3) + (0.2)*(y(4){-}y(3))$$

and

$$alty3pt2 = sin(2*pi*2.2 / 511)$$

which yield *y3pt2* = 0.02699 and *alty3pt2* = 0.02704, which differ by about 5.3 parts in one hundred thousand. The error varies according to where in the cycle of the sinusoid you are interpolating.

The script (see exercises below)

$$LVxSineROMNonIntDecInterp(N, F\_ROM, D)$$

affords experimentation with nonintegral decimation of a sinusoid. For this script, D may have a nonintegral value. A typical call, which results in Fig. 3.39, would be

LVxSineROMNonIntDecInterp(32,1,2.5)

In plot (a) of Fig. 3.39, the known (ROM) sample values are plotted as stems with small circles, and the values computed by linear interpolation for a decimation factor of 2.5 (sample index numbers 1, 3.5, 6, 8.5, etc.) are plotted as stems with diamond heads. In the lower plot, the computed values marked with diamonds in the upper plot are extracted and plotted to show the decimated output sequence (the computation shown in the figure was stopped before the computed value of the address exceeded the ROM length in order to keep the displayed mixture of existing ROM samples and interpolated samples from becoming confusing).

The decimation factor may also be less than 1.0, which leads to a frequency reduction rather than a frequency increase. Figure 3.40 shows the first 32 samples of the decimated output sequence when the decimation factor is set at 0.5.

3.18   COMPRESSION

An interesting and useful way to reduce the apparent quantization noise when using a small number of bits (such as eight or fewer) is to use A-Law or μ-Law compression. These two schemes, which are very similar, compress an audio signal in analog form prior to digitization.

In an analog system, where μ-Law compression began, small signal amplitudes are boosted according to a logarithmic formula. The signal is then transmitted through a noisy channel. Since the (originally) lower level audio signals, during transmission, have a relatively high level, their signal to noise ratio emerging from the channel at the receiving end is much better than without compression.

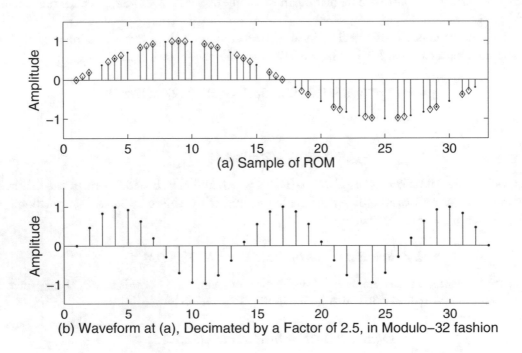

Figure 3.39: (a) One cycle of a 32-sample sinusoid, interpolated values (every 2.5 th sample) marked with diamonds; (b) Net decimated output sequence, formed by extracting the interpolated samples (marked with diamonds) from the upper plot.

The expansion process adjusts the gain of the signal inversely to the compression characteristic, but the improved signal-to-noise ratio remains after expansion.

Using a sampled signal sequence, the signal samples have their amplitudes adjusted (prior to being quantized) according to the following formula:

$$F(s[n]) = S_{MAX}(\log(1 + \mu \frac{s[n]}{S_{MAX}})/\log(1 + \mu))sgn(s[n]) \qquad (3.6)$$

where S_{MAX} is the largest magnitude in the input signal sequence $s[n]$, sgn is the sign function, $F(s[n])$ is the compressed output sequence, and μ is a parameter which is typically chosen as 255.

In compression, smaller signal values are boosted significantly so that there are far fewer small values hovering near zero amplitude. This is because with a small number of bits, small amplitudes suffer considerably more than larger amplitudes. Pyschoacoustically speaking, larger amplitude sounds mask noise better than smaller amplitude sounds. A large step-size (and the attendant quantization noise) at low signal levels is much more audible than the same step-size employed at high amplitude levels. Additionally, small amplitudes which would be adequately encoded using a large

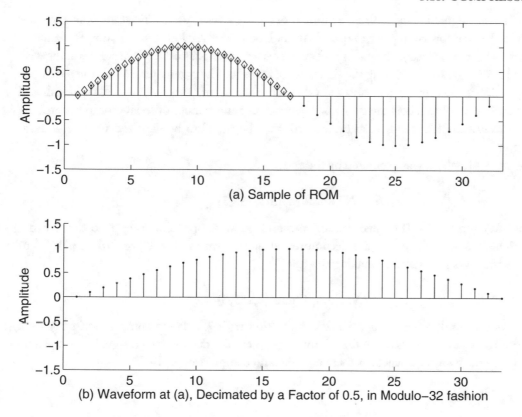

Figure 3.40: (a) One cycle of a 32-sample sinusoid, interpolated values (every half sample) marked with diamonds; (b) Net decimated output sequence, formed by extracting the interpolated samples (marked with diamonds) from the upper plot.

number of bits may actually be encoded as zero when using a small number of bits, leading to a squelch effect which can be disconcerting.

In μ-Law compression, the interesting result is that smaller signal amplitudes wind up being quantized (after reconstruction) with smaller amplitude differences between adjacent quantized levels than do larger signal amplitudes. As a result, when the signal amplitude is low, accompanying quantization noise is also less, and thus is better masked.

The script (see exercises below)

LVxCompQuant(TypeCompr,AMu,NoBits)

affords experimentation with μ-law compression as applied or not applied to an audio signal quantized to a specified number of bits. Passing *TypeCompr* as 1 results in no μ-law compression being applied, the audio signal is simply quantized to the specified number of bits and then converted back to an analog signal having a discrete number of analog levels. Passing *TypeCompr* as 2 invokes the use of μ-law compression prior to quantization. Upon reconversion to the analog domain and decompression (or expansion), the spacing between adjacent quantization levels decreases along with sample amplitude, greatly improving the signal-to-noise ratio. The parameter *AMu* is the μ-law parameter and 255 is the most commonly used value. The number of quantization bits, *NoBits*, should be passed as 2 or 3 to best show the difference between use and nonuse of μ-law compression. The audio file *drwatsonSR8K.wav* is used by the script as the signal to be digitized with or without compression.

Figure 3.41, which was generated by the call

$$\textbf{LVxCompQuant(1,255,3)}$$

shows the situation for 3 bits of quantization and no μ−law compression as applied to the audio file *drwatsonSR8K.wav*. Compare the decompressed file, shown in plot (c), to that of plot (c) of Fig. 3.42, which was generated by making the call

$$\textbf{LVxCompQuant(2,255,3)}$$

which shows the result when using μ−law compression with 255 as the compression parameter. Note in the latter case that there are few if any cases where the decompressed signal is completely zeroed out or effectively squelched, as there are when no compression is used.

3.19 REFERENCES

[1] James H. McClellan, Ronald W. Schaefer, and Mark A. Yoder, *Signal Processing First*, Pearson Prentice Hall, Upper Saddle River, New Jersey, 2003.

[2] Alan V. Oppenheim and Ronald W. Schaefer, *Discrete-Time Signal Processing*, Prentice-Hall, Englewood Cliffs, New Jersey, 1989.

[3] John G. Proakis and Dimitris G. Manolakis, *Digital Signal Processing, Principles, Algorithms, and Applications, Third Edition*, Prentice Hall, Upper Saddle River, New Jersey, 1996.

[4] Ken C. Pohlmann, *Principles of Digital Audio*, Second Edition, SAMS, Carmel, Indiana.

[5] John Watkinson, *The Art of Digital Audio, First Edition*, (Revised Reprint, 1991), Focal Press, Jordan Hill, Oxford, United Kingdom.

[6] James A. Blackburn, *Modern Instrumentation for Scientists and Engineers*, Springer-Verlag, New York.

[7] Chris Dick and Fred Harris, "FPGA Interpolators Using Polynomial Filters," *The 8th International Conference on Signal Processing Applications and Technology, Toronto, Canada*, September 11, 1998.

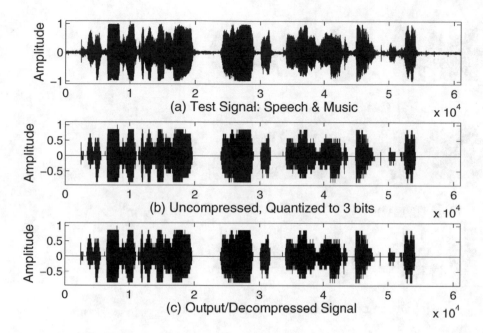

Figure 3.41: (a) Test signal (analog); (b) Test signal, without prior compression, quantized to 3 bits; (c) Analog signal reconstructed from quantized samples at (b), exhibiting a squelch effect, or total loss of reconstructed signal when the original signal level was very low.

[8] Valimaki and Laakso, "Principles of Fractional Delay Buffers," *IEEE International Conference on Acoustics, Speech, and Signal Processing (ICASSP'00), Istanbul, Turkey*, 5-9 June 2000.

[9] Mar et al, "A High Quality, Energy Optimized, Real-Time Sampling Rate Conversion Library for the StrongARM Microprocessor," *Mobile & Media Systems Laboratory, HP Laboratories Palo Alto*, HPL-2002-159, June 3, 2002.

3.20 EXERCISES

1. An ADC that operates at a rate of 10 kHz is not preceded by an anti-aliasing filter. Construct an appropriate folding diagram and give the apparent output frequency and phase (in-phase or phase-reversed) corresponding to each of the following input frequencies:
   - (a) 4 kHz
   - (b) 5 kHz
   - (c) 9 kHz
   - (d) 10 kHz
   - (e) 16 kHz

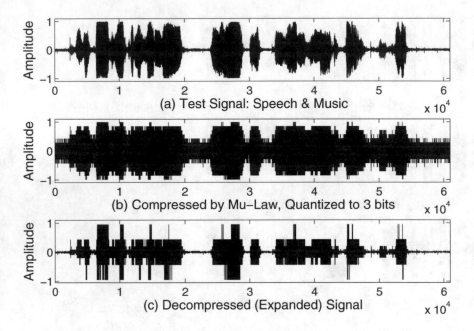

Figure 3.42: (a) Test signal (analog); (b) Quantized version of test signal, which was compressed using μ−law compression prior to being quantized; (c) Analog signal reconstructed from quantized samples at (b) by first converting from quantized form to analog, then decompressing the analog signal values. Note that the compression technique has prevented the squelch effect from occurring.

(f) 47 kHz

(g) 31 kHz

2. For each of the input frequencies given above, and using a sampling rate of 10 kHz, generate and plot the ADC's sequence of output values. To do this, set up a time vector with one second's worth of sampling times, and evaluate the values of a sine or cosine for the given input frequency, and plot the result. Note: it is helpful to plot only about the first 50 samples of the output rather than the entire output.

3. State what value of cutoff frequency an anti-aliasing filter should have for the ADC of the above exercises (i.e., an ADC operating at sampling rate of 10 kHz).

4. A certain system transmits any one of eight frequencies which are received at the input of an ADC operating at 10 kHz. An anti-aliasing filter cutting off above 10 kHz (not 5 kHz, the Nyquist rate) precedes the ADC. Using the folding diagram for 10 kHz, state the apparent frequency in the output corresponding to each of the following eight input frequencies: [1 kHz, 2kHz, 3kHz, 4 kHz, 6.5 kHz, 7.5 kHz, 8.5 kHz, and 9.5 kHz.

5. MathScript provides various functions that return b and a coefficients that can be used to implement filters of various types, such as lowpass, highpass, bandpass, and notch. The frequency arguments for such functions are given in normalized frequency. For example, the function

$$b = fir1(N, Wn)$$

where N is the filter order (one fewer than the filter or impulse response length that will be provided as b), and Wn is the normalized frequency specification. Lowpass and highpass filter impulse responses that have cutoffs at one-half the Nyquist rate, for example, and a length of 21 samples (for example), can be obtained by making the call

$$b = fir1(20,0.5)$$

for the lowpass filter, and

$$b = fir1(20,0.5,'high')$$

for the highpass filter.

    Likewise, bandpass or notch filters of length 31, for example, and band limits of 0.4 and 0.6 (for example) can be obtained with the call

$$b = fir1(30,[0.4,0.6])$$

for the bandpass filter and

$$b = fir1(30,[0.4,0.6],'stop')$$

for the notch filter.

    For the following sequence sample rates, desired filter length, and desired actual frequency cutoff limits, design the call for the $fir1$ function; verify the filter function by filtering a chirp of frequency range 0 to the respective Nyquist rate with the impulse response by using the MathScript function *filter*.

    (a) SR = 25,000; filter length = 37; [7000] (lowpass)
    (b) SR = 35,000; filter length = 47; [11000] (lowpass)
    (c) SR = 44,100; filter length = 107; [5000,8000] (bandpass)
    (d) SR = 48,000; filter length = 99; [65] (high)
    (e) SR = 8,000; filter length = 47; [800,1200] (notch)

6. What normalized frequency will generate a complex exponential exhibiting one cycle for every

    (a) 2 samples
    (b) 4 samples
    (c) 8 samples
    (d) 17.5 samples

(e) 3.75 samples

For each of (a) through (e), devise a call to generate and plot the real and imaginary parts of a complex exponential having the respective stated period.

7. Using paper and pencil, convert the following decimal numbers to binary offset notation using 4 bits:

(a) 0; (b) 1; (c) 2; (d) 3; (e) 8; (f) 9; (g) 15

8. Using paper and pencil, convert the following binary numbers (in offset format) to decimal equivalent:

(a) [1001001]; (b) [10000001]; (c) [10101010]; (d) [11111111]

9. Write a script that implements the following function:

$$[\text{OutputMat}] = \text{LVxDigitizePosNums(NumsToDig,NumBits)}$$

in which *OutputMat* is a matrix of binary numbers, each row of which represents one number from the input vector *NumsToDig*, which consists of zero or positive decimal numbers to be converted to binary offset format, and NumBits is the minimum number of bits necessary to quantize all the decimal numbers in *NumsToDig* without clipping.

function [OutputMat] = LVxDigitizePosNums(xNumsToDig,nBits)
% Receives a vector of positive decimal numbers to convert to
% binary notation using nBits number of bits, and produces
% as Output a matrix each row of which is a binary
% representation of the corresponding input decimal number.
% Test call:
% LVxDigitizePosNums([0,1,34,23,2,17,254,255,127],8)

The following steps need to be completed:

(a) From the input arguments, generate a power-of-two vector of appropriate length;

(b) Initialize *OutputMat* as a matrix of zeros having a number of rows equal to the length of vector *NumsToDig*, and a number of columns equal to *NumBits*;

(c) To conduct the actual successive approximation algorithm, all the values in one column of *OutputMat* at a time are set to 1, and the following matrix equation is evaluated:

$$\text{TestMat} = \text{OutputMat*xWtVec' - NumsToDig';}$$

(d) The row indices must be found of entries in *TestMat* (which is a column vector) that are less than zero, and the corresponding values in *OutputMat* are then reset from 1 to zero.

(e) The operation proceeds in this manner from column to column until all columns have been experimentally set to 1, tested, and reset to zero as required. The final result is then the output of the function, *OutputMat*.

10. Write a script that can receive decimal numbers from 0.00 up to 10.00 and convert them to binary offset notation having 16 bits of precision, using the method of successive approximation. State the value of the LSB.

11. A certain audio signal has values ranging from -50 to +50 volts, and is to be quantized using an ADC that accepts an input voltage range of 0 to +10 volts. It is desired to quantize the audio signal to an accuracy of at least one part per thousand. Describe the signal adjustments necessary to make the audio signal suitable as an input signal, and determine the minimum number of bits of precision needed in the ADC. State the value of the resulting LSB in volts. What are the minimum and maximum binary output values from the ADC assuming offset notation?

12. Write a script that uses the provided wave file *drwatsonSR8K.wav* as a test signal, compresses it according to μ-Law compression as described above using the input parameter *AMu* as the compression parameter μ, quantizes it using a user-selectable number of bits *NoBits*, decompresses it, and allows you to play it through your computer's sound card.

LVxCompQuant(TypeCompr,AMu,NoBits)

For comparison purposes, you should be able to process and play the test signal with and without use of μ-Law compression according to the value of the input argument *TypeCompr*.

To perform the decompression, first generate the decompression formula by solving for $s[n]$ in (3.6); a helpful thing to do at the signal creation stage is to divide the signal by the maximum of its absolute value so that $S_{MAX} = 1$, thus simplifying the algebra in the decompression formula.

Vectors may be played using the function

sound(AudioSig,NetSR)

where *AudioSig* is the vector, *NetSR* is the sample rate at which it to be played. *NetSR* must be any of 8 kHz, 11.025 kHz, 22.05 kHz, or 44.1 kHz.

13. Write a script that conforms to the following call syntax

BinaryMatrix = LVxADCPosNegQuants(NumsToDig)

where *NumsToDig is* a decimal number array to be converted to a binary array, each binary number in the array having a single sign bit followed by N bits representing the magnitude, for a total of N + 1 bits per binary number. Positive decimal numbers are converted to binary representations and a sign bit of 0 is placed to the left of the magnitude's MSB. Negative numbers have their magnitudes converted to binary representations and a sign bit of 1 is used.

The input argument consists of a vector of real, positive and/or negative decimal numbers, and the function returns a matrix of binary numbers, each original decimal number occupying one row of the returned matrix. The function automatically determines how many bits will be required, which depends on the number with the largest magnitude in the vector of numbers to be digitized (an LSB value of 1.0 is assumed for simplification). For example, the call

LVxADCPosNegQuants([17,7,-100,77])

should return the following binary matrix:

$$\begin{bmatrix} 0 & 0 & 0 & 1 & 0 & 0 & 0 & 1 \\ 0 & 0 & 0 & 0 & 0 & 1 & 1 & 1 \\ 1 & 1 & 1 & 0 & 0 & 1 & 0 & 0 \\ 0 & 1 & 0 & 0 & 1 & 1 & 0 & 1 \end{bmatrix}$$

in which the first row is the binary representation of the first number in the vector of decimal numbers to be converted (17 in this case), and so forth. The leftmost bit of each row is the sign bit; positive numbers have "0" for a sign bit, and negative numbers have a "1."

14. Write a script that conforms to the following call syntax

$$BinOut = LVxBinaryCodeMethods(BitsQ, SR, Bias, Freq, Amp, CM, PlotType)$$

as described in the text and below:

> **function BinOut = LVxBinaryCodeMethods(BitsQ,SR,Bias,Freq,...**
> **Amp,CodeMeth,PlotType)**
> **% Quantizes a sine wave of amplitude Amp and frequency**
> **% Freq using BitsQ number of quantization bits at a sample**
> **% rate of SR.**
> **% Bias: 0 for none, 1 for 1/2 LSB**
> **% CodeMeth: 1 = Sign + Mag; 2 = Offset;**
> **% PlotType: val 0 plot as multiples of LSB, or 1 to plot in volts**
> **% Generates a figure with three subplots, the first is the**
> **% (simulated) analog test signal, the second has an overlay**
> **% of the simulated test signal and its quantized version, and the**
> **% third is the quantization error**
> **% Test calls:**
> **% BinOut = LVxBinaryCodeMethods(4, 1000,0, 10,170,2,0)**
> **% BinOut = LVxBinaryCodeMethods(2, 1000,0, 10,170,1,1)**
> **% BinOut = LVxBinaryCodeMethods(3, 1000,1, 10,170,1,1)**

15. Write the m-code for the script

$$LVxInterpolationViaSinc(N, SampDecRate, valTestSig)$$

which conforms to the syntax below, functions as described in the text, and which creates plots as shown in the text. Test it with the given sample calls:

> **function LVxInterpolationViaSinc(N,SampDecRate,valTestSig)**

% N is the master sequence length, from which a densely-sampled
% test signal is generated. The master sequence is decimated by
% every SampDecRate samples to create the test sample
% sequence from which an interpolated version of the underlying
% bandlimited continuous domain signal will be generated.
% N must be an even integer, and SampDecRate must be an
% integer; valTestSig may be any integer from 1 to 5, such that
% 1 yields the waveform cos(2*pi*n*0.1) + 0.7*sin(2*pi*n*0.24);
% 2 yields DC
% 3 gives a single triangular waveform
% 4 gives two cycles of a triangular waveform
% 5 gives 0.5*cos(2*pi*n*0.125) + 1*sin(2*pi*n*0.22);
% where n = -N/2:1:N/2 if N is even
% Sample calls:
% LV_InterpolationViaSinc(1000,100,1)
% LV_InterpolationViaSinc(1000,50,1)
% LV_InterpolationViaSinc(1000,25,1)
% LV_InterpolationViaSinc(5000,100,1)

16. Write the m-code for the script

$$LVxSineROMNonIntDecInterp(N, F\_ROM, D)$$

as described and illustrated in the text. Test the script with the following calls:

   (a) **LVxSineROMNonIntDecInterp(77,1,7.7)**
   (b) **LVxSineROMNonIntDecInterp(23,2.5,0.6)**
   (c) **LVxSineROMNonIntDecInterp(48,3,4)**

17. A chirped audio signal which is of one second duration and which sweeps from 0 Hz to 8000 Hz in one second is sampled at 2000 Hz. Compute and plot the resultant wave State at what times during the one second interval maxima and minima in observed (or audible or apparent) frequency occur.

18. Write the m-code for the following function, which uses sinc interpolation to convert an audio file sampled at 8 kHz to one sampled at 11.025 kHz.

   function LVxInterp8Kto11025(testSigType,tsFreq)
   % Converts an audio signal having a sample rate of 8 kHz to
   % one sampled at 11.025 kHz using both sinc and linear
   % interpolation. The audio files used can be either the audio
   % file 'drwatsonSR8K.wav' or a cosine of user designated
   % frequency. For sinc interpolation it uses 10 samples of the
   % input signal and a 100 by 10 sinc interpolation matrix to

```
% generate sample values located at the fractional sample
% index values 1:320/441:length(OriginalAudioFile)
% To use 'drwatsonSR8K.wav', pass testSigType as 1 and
% tsFreq as [] or any number; to use a cosine, pass
% testSigType as 0 and pass tsFreq as the desired cosine
% frequency in Hz. The script automatically plays the original,
% sinc-interpolated, and linear interpolated signals using the
% Mathscript function sound., and plots, for a short sample
% interval, the original, sinc-interpolated, and linear interpolated
% signal values on one axis for comparison prior to
% post-interpolation lowpass filtering, which limits the
% passband of the interpolated signals to the original 4 kHz
% bandwidth.
% Test calls:
% LVxInterp8Kto11025(0,250)
% LVxInterp8Kto11025(0,2450)
% LVxInterp8Kto11025(1,[])
```

The sinc interpolation matrix may, for example, consist of rows 900-999 as seen in Fig. 3.31. As an example computation, if (say) a sample is needed at index 10.71, the ten columns of row 72 (the row index must start with one rather than zero) of the sinc matrix would be multiplied by original signal samples 6-15, respectively, with sample 10 (i.e., **floor(10.71)**) being multiplied by column 5 as shown in Fig. 3.31, and sample 11 (i.e., **ceil(10.71)**) being multiplied by column 6. The products are summed to obtain the interpolated value. The operation is actually the inner or dot product of the chosen sinc matrix row with a column vector selected from the signal to be interpolated.

It is only necessary to perform the interpolation for a limited number of samples to demonstrate efficacy; 1000-2000 samples are enough at a sample rate of 8kHz to hear that the original and interpolated versions have the same apparent frequency.

The interpolated samples should be lowpass filtered after interpolation to retain the original frequency limit, which is 4 kHz. The following code may be used:

```
[b,a] = cheby1(12,0.5,7.9/11.025);
SincInterpSig = filter(b,a,SincInterpSig);
LinInterpSig = filter(b,a,LinInterpSig);
```

CHAPTER 4

Transform and Filtering Principles

4.1 OVERVIEW

Having become acquainted in the last chapter with basic signal acquisition (ADC) and reconstruction techniques (DAC) and the all-important concepts of Nyquist rate and normalized frequency, we are now in a position to investigate the powerful principle of **Correlation**. Discrete frequency transforms, which can be used to determine the frequency content or response of a discrete signal, and time domain digital filters (i.e., the FIR and IIR), which can preferentially select or reject certain frequencies in a signal, function according to two principles of correlation—namely, respectively, the single-valued correlation of two equally-sized, overlappingly aligned waveforms, and the **Correlation Sequence**. Another principle, that of orthogonality, also plays an important role in frequency transforms, and this too will be explored to show why frequency transforms such as the DFT require that two correlations (or a single complex correlation) be performed for each frequency being tested.

We begin our discussion with an elementary concept of correlation, the correlation of two samples, and quickly move, step-by-step, to the correlation of a signal of unknown frequency content with two equally-sized orthogonal sinusoids (i.e., cosine-sine pairs) and, in short order, the real DFT. We then briefly illustrate the use of the property of orthogonality in signal transmission, a very interesting topic which illustrates the power of mathematics to allow intermixing and subsequent decoding of intelligence signals. From there we investigate the correlation sequence, performing correlation via convolution, and matched filtering. We then informally examine the frequency selective properties of the correlation sequence, and learn how to construct basic (although inefficient) filters. We learn the principle of sinusoidal fidelity and then determination of time delays between sequences using correlation, an application of correlation that is often used in echo canceller training and the like. For the final portion of the chapter, we investigate simple one- and two-pole IIRs with respect to stability and frequency response, and demonstrate how to generate IIRs having real-only filter coefficients by using poles in complex conjugate pairs.

By the end of this chapter, the reader will be prepared to undertake the study of discrete frequency transforms as found in Volume II in this series, which includes a general discussion of various transforms and detailed chapters on the Discrete Time Fourier Transform (DTFT), the Discrete Fourier Transform (DFT), and the z-Transform.

4.2   SOFTWARE FOR USE WITH THIS BOOK

The software files needed for use with this book (consisting of m-code (.m) files, VI files (.vi), and related support files) are available for download from the following website:

http://www.morganclaypool.com/page/isen

The entire software package should be stored in a single folder on the user's computer, and the full file name of the folder must be placed on the MATLAB or LabVIEW search path in accordance with the instructions provided by the respective software vendor (in case you have encountered this notice before, which is repeated for convenience in each chapter of the book, the software download only needs to be done once, as files for the entire series of four volumes are all contained in the one downloadable folder).

See Appendix A for more information.

4.3   CORRELATION AT THE ZEROTH LAG (CZL)

A simple concept of numerical correlation is this: if two numbers are both positive, or both negative, they correlate well; if one is positive, and one is negative, they correlate strongly in a negative or opposite sense. A way to quantify this idea of correlation is simply to multiply the two numbers. If they are both positive, or both negative, the product is positive; if one number is positive, and one number is negative, the product is negative. Thus two numbers having the same sign have a positive correlation value, while two numbers having opposite signs have a negative correlation value. If one number is zero, the product is zero, and no correlation between the two numbers can be determined, that is, they are neither alike nor unalike; in this case they are termed "uncorrelated."

To correlate two waveforms (i.e., sequences) having the same length N, multiply corresponding samples and add up all the products. Imagine this as first laying one waveform on top of the other so that the first sample of one waveform lies atop the first sample of the other, and so forth. Then multiply each pair of corresponding samples and add the products.

Stated mathematically, the correlation of two sequences $A[n]$ and $B[n]$ having equal lengths N is

$$CZL = \sum_{n=0}^{N-1} A[n]B[n] \tag{4.1}$$

where CZL is the **Correlation at the Zeroth Lag** of sequences $A[n]$ and $B[n]$. The nomenclature *Zeroth Lag* is used in this book for clarity to denote this particular case, in which two sequences of equal length are correlated with no offset or delay (lag) relative to one another. In most books, the process represented by Eq. (4.1) is usually referred to simply as the "correlation" of sequences $A[n]$ and $B[n]$.

Example 4.1.   Compute the CZL of the sequence $[ones(1, 16), zeros(1, 16)]$ with itself. Follow up by correlating the sequence with the negative of itself.

Figure 4.1, plots (a) and (b), show two instances of the subject 32-sample sequence. To compute the correlation value, figuratively lay the first waveform on top of the second. Multiply all overlapping samples, and add the products to get the answer, which is 16. Since the two waveforms of Fig. 4.1 are in fact the same, they have a high positive value of correlation, which is expected from the basic concept of correlation.

The sequence and its negative are shown in plots (c) and (d) of Fig. 4.1. The correlation value in this case is −16, indicating a strong anti-correlation.

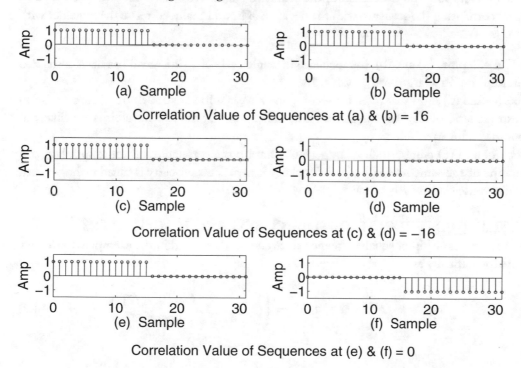

Figure 4.1: (a) First sequence; (b) Second sequence, identical to the 1st, leading to a large positive correlation; (c) First sequence; (d) Second sequence, the negative of the 1st, leading to a large negative correlation; (e) First sequence; (f) Second sequence, having no overlapping sample pairs with the first sequence that have nonzero products, leading to a correlation value of 0.

Example 4.2.   Compute the CZL of the two sequences [ones(1, 16), zeros(1, 16)] and [zeros(1, 16), -ones(1, 16)].

The two sequences are shown in plots (e) and (f) of Fig. 4.1. The correlation value is zero since (when you figuratively lay one waveform atop the other) one value of each pair of overlapping samples is always zero, yielding a product of zero for all 32 sample pairs, summing to zero.

Example 4.3.   Construct two sequences, two different ways, of length eight samples, using as values only 1 or -1, which yield a CZL of 0.

Two such possible sequences are: First pair: [-ones(1,8)] and [-ones(1,4), ones(1,4)]; a second pair might be: [-ones(1,8)] and [-1,-1,1,1,-1,-1,1,1]. The reader should be able to construct many more examples. For example, if, of the first pair, the first vector were [ones(1,8)] instead of [-ones(1,8)], the CZL would still be zero.

Example 4.4.   Compute the correlation of the sequence $\sin[2\pi n/32]$ with itself ($n = 0 : 1 : 31$). Follow up by correlating the sequence with the negative of itself. Finally, correlate the sequence with $\cos[2\pi n/32]$.

In Fig. 4.2, plots (a) and (b), the sequence (a single cycle of a sine wave) has been correlated with itself; resulting in a large positive value, 16.

Plots (c) and (d) show that when the second sine wave is shifted by 180 degrees, the correlation value, as expected, becomes a large negative value, indicating that the two waveforms are alike, but in the opposite or inverted sense.

Plots (e) and (f) demonstrate an important concept: the correlation (at the zeroth lag) of a sine and cosine of the same integral-valued frequency is zero. This property is called orthogonality, which we'll consider extensively below.

4.3.1   CZL EQUAL-FREQUENCY SINE/COSINE ORTHOGONALITY

The correlative relationship of equal-frequency sines and cosines noted in the examples above may be stated mathematically as

$$\sum_{n=0}^{N-1} \sin(2\pi kn/N) \sin(2\pi kn/N) = \begin{cases} 0 & \text{if} \quad k = [0, N/2] \\ N/2 & \text{if} \quad \text{otherwise} \end{cases} \tag{4.2}$$

and

$$\sum_{n=0}^{N-1} \sin(2\pi kn/N) \cos(2\pi kn/N) = 0 \tag{4.3}$$

where k is an integer, N is the sequence length, and n = 0:1:$N - 1$. The latter equation expresses the principle of orthogonality of equal integral-frequency sine and cosine waves, which is simply

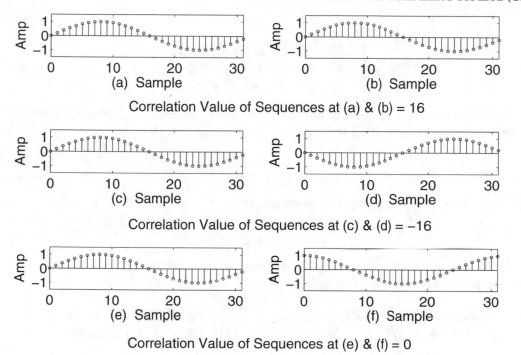

Figure 4.2: (a) First sequence, a sine wave; (b) Second sequence, identical to the first, yielding a large positive correlation; (c) First sequence; (d) Second sequence, the negative of the first, showing a large negative correlation; (e) First sequence; (f) Second sequence, a cosine wave, yielding a correlation value of 0.

that their correlation value over one period is zero. When cosines are used instead of sines, the relationship is

$$\sum_{n=0}^{N-1} \cos(2\pi kn/N)\cos(2\pi kn/N) = \begin{cases} N & \text{if} \quad k = [0, N/2] \\ N/2 & \text{if} \quad \text{otherwise} \end{cases}$$

4.3.2   CZL OF SINUSOID PAIRS, ARBITRARY FREQUENCIES

Consider the following sum S_C where N is the number of samples in the sequence, n is the sample index, and k_1 and k_2 represent integer frequencies such as -3, 0,1, 2, etc.

$$S_C = \sum_{n=0}^{N-1} \cos[2\pi k_1 n/N]\cos[2\pi k_2 n/N] \tag{4.4}$$

yields

$$S_C = \begin{cases} N/2 & \text{if} \quad [k_1 = k_2] \neq [0, N/2] \\ N & \text{if} \quad [k_1 = k_2] = [0, N/2] \\ 0 & \text{if} \quad k_1 \neq k_2 \end{cases} \quad (4.5)$$

- In the statements above, it should be noted that values for frequencies k_1, k_2 should be understood as being modulo N. That is to say, the statement (for example)

$$[k_1 = k_2] \neq 0$$

as well as all similar statements, should be interpreted as

$$[k_1 = k_2] \neq 0 \pm mN$$

where m is any integer ...–2,1,0,1,2...or in plain language as "k_1 is equal to k_2, but k_1 (and k_2) are not equal to 0 plus or minus any integral multiple of N."

Example 4.5.   Compute the correlation (CZL) of two cosines of the following frequencies, having $N = 16$: $a)$ [1, 1]; $b)$ [0, 0]; $c)$ [8, 8]; $d)$ [6, 7]; $e)$ [1, 17]; $f)$ [0, 16]; $g)$ [8, 40]; $h)$ [6, 55].

To make the computations easy, we provide a simple function:

function [CorC] = LVCorrCosinesZerothLag(k1,k2,N)
n = 0:1:N-1;
CorC = sum(cos(2*pi*n*k1/N).*cos(2*pi*n*k2/N));

and after making the appropriate calls, we get the following answers:

a) In this case, $k_1 = k_2 \neq [0, 8]$, yielding $N/2 = 8$ in this case
b) In this case, $k_1 = k_2 = 0$, yielding $N = 16$
c) In this case, $k_1 = k_2 = N/2$, yielding $N = 16$
d) In this case, $k_1 \neq k_2$ yielding 0
e) In this case, 17 $(1+N)$ is equivalent to 1; the call

$$[\text{CorC}] = \text{LVCorrCosinesZerothLag(1,17,16)}$$

yields $CorC = 8$, just as for case (a) in which the frequencies were [1,1].
    f) Note that 16 is $(0 +N)$, so this is equivalent to case (b) and the call

$$[\text{CorC}] = \text{LVCorrCosinesZerothLag(0,16,16)}$$

yields $CorC = 16$, just as for case (b) in which the frequencies were [0,0].
    g) Note that the second frequency, 40, is equal to $(8+2N)$, so this is equivalent to case (c) in which the frequencies were [8,8]; the call

$$[CorC] = \textbf{LVCorrCosinesZerothLag(8,40,16)}$$

yields $CorC = 16$, just as for case (c) in which the frequencies were [8,8].

h) The frequency 55 is equivalent to (7+3N), so this is equivalent to case (d); the call

$$[CorC] = \textbf{LVCorrCosinesZerothLag(6,55,16)}$$

yields $CorC = 0$, just as for case (d) in which the frequencies were [6,7].

The CZL of two sine waves having arbitrary, integral-valued frequencies, defined as

$$S_S = \sum_{n=0}^{N-1} \sin(2\pi k_1 n/N)\sin(2\pi k_2 n/N)$$

yields

$$S_S = \begin{cases} N/2 & \text{if} \quad [k_1 = k_2] \neq [0, N/2] \\ 0 & \text{if} \quad [k_1 = k_2] = [0, N/2] \\ 0 & \text{if} \quad k_1 \neq k_2 \end{cases} \tag{4.6}$$

where k_1 and k_2 are integers and are modulo-N as described and illustrated above.

Figure 4.3 shows the values of S_C and S_S for the case of $k1 = 0$ with $k2$ varying from 0 to 5, while Fig. 4.4 shows S_C and S_S for the case of $k1 = 3$ with $k2$ varying from 0 to 5.

Example 4.6. Compute the CZL of two sine waves having frequencies of one and two cycles, respectively, over 32 samples.

Figure 4.5 shows one and two cycle sine waves at plots (a) and (b) respectively. Since the two sine waves differ in frequency by an integer, the CZL is zero, as displayed beneath the plots.

Plots (b) and (c) show another example in which the two sine waves have frequencies of three and five cycles over the sequence length, with the predicted CZL of 0.

Figure 4.5, plot (c), shows the CZL of a four cycle sine wave and one of 5.8 cycles. In this case, the CZL is not guaranteed to be zero, and in fact it is not equal to zero.

4.3.3 ORTHOGONALITY OF COMPLEX EXPONENTIALS

The sum of the product (i.e., the CZL) of two complex exponentials each having an integral number of cycles k_1, k_2, respectively, over the sequence length N, where k_1 and k_2 are modulo-N, is

$$\sum_{n=0}^{N-1} e^{j2\pi k_1 n/N} e^{-j2\pi k_2 n/N} = \begin{cases} 0 & \text{if} \quad k_1 \neq k_2 \\ N & \text{if} \quad \text{otherwise} \end{cases}$$

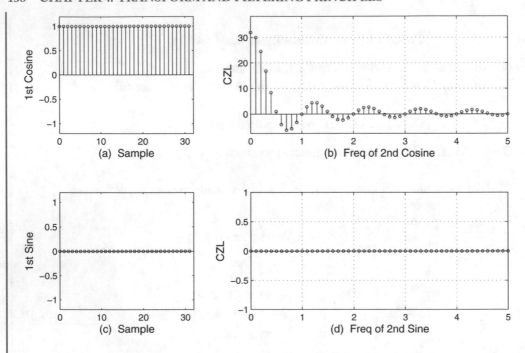

Figure 4.3: (a) First cosine, $k1 = 0$; (b) CZL between first and second cosines, the second cosine frequency $k2$ varying between 0 and 5 cycles per 32 samples; (c) First sine, $k1 = 0$; (d) CZL between first and second sines, the second sine frequency $k2$ varying between 0 and 5 cycles per 32 samples.

4.3.4   SUM OF SAMPLES OF SINGLE COMPLEX EXPONENTIAL

An interesting and useful observation is that if one of the complex exponentials has its frequency as zero, it is identically equal to 1.0, and the sum reduces to

$$\sum_{n=0}^{N-1} e^{\pm j2\pi k_1 n/N} = \begin{cases} N & \text{if} \quad k_1 = \ldots - N, 0, N, 2N, \ldots \\ 0 & \text{if} \qquad\qquad \text{otherwise} \end{cases}$$

For the sum of the samples of a single complex exponential, use the following:

function s = LVSumCE(k,N)
% s = LVSumCE(2,32)
n = 0:1:N-1; s = sum(exp(j*2*pi*n*k/N))

The call

$$s = \textbf{LVSumCE(2,32)}$$

for example, yields the sum as s = -3.3307e-016, which differs from zero only by reason of computer roundoff error, whereas a call in which the frequency argument is equal to N, such as

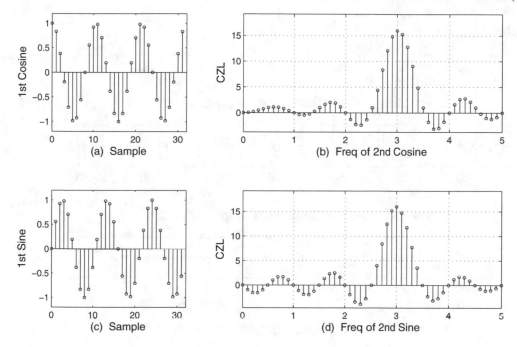

Figure 4.4: (a) First cosine, $k1 = 3$; (b) CZL between first and second cosines, the second cosine frequency $k2$ varying between 0 and 5 cycles per 32 samples; (c) First sine, $k1 = 3$; (d) CZL between first and second sines, the second sine frequency $k2$ varying between 0 and 5 cycles per 32 samples.

$$s = LVSumCE(32,32)$$

yields $s = 32$.

4.3.5   IDENTIFYING SPECIFIC SINUSOIDS IN A SIGNAL

We now consider the problem of identifying the presence or absence of sinusoids of specific frequency and phase in a signal. We begin the discussion with a specific problem set forth in the following example.

Example 4.7.   A sequence $A[n]$ can assume the values of 0, $\sin[2\pi n/N + \theta]$, or $\sin[4\pi n/N + \theta]$ where n is the sample index, N is the sequence length, θ is a variable phase which may assume the values 0, $\pi/4$, $\pi/2$, $3\pi/4$, or π radians. Devise a way to determine the signal present in $A[n]$ using correlation.

Let's consider several cases to see what the difficulties are:

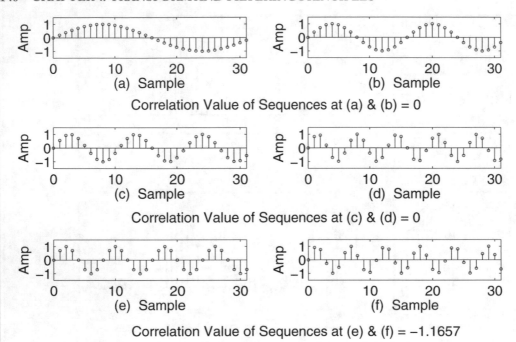

Correlation Value of Sequences at (a) & (b) = 0

Correlation Value of Sequences at (c) & (d) = 0

Correlation Value of Sequences at (e) & (f) = −1.1657

Figure 4.5: (a) One cycle of a sine wave; (b) Two cycles of a sine wave, with correlation at the zeroth lag (CZL) value of zero; (c) Three cycles of a sine wave; (d) Five cycles of a sine wave; (e) Four cycles of a sine wave; (f) Five point eight (5.8) cycles of a sine wave.

Suppose in one case that $A[n] = \sin[2\pi n/N + \theta]$ and $\theta = 0$. The correlation of $A[n]$ with a test sine wave $T\,S[n] = \sin[2\pi n/N]$ will yield a large positive correlation value. If, however, $\theta = \pi/2$ radians (90 degrees), then the same correlation would yield a value of zero due to orthogonality.

In another case, suppose $A[n] = 0$. The correlation of $A[n]$ with $T\,S[n]$ yields zero irrespective of the value of θ. Unfortunately, we don't know if that means the unknown signal is simply identically zero (at least at samples where the test sine wave is nonzero), or if the unknown signal happens to be a sinusoid that is 90 degrees out of phase with the test sine wave $T S[n]$.

In yet a third case, if $A[n] = \sin[4\pi n/N + \theta]$, any correlation with $T\,S[n]$ will also yield zero for any value of θ.

The solution to the problem, then, is to perform two correlations for each test frequency with $A[n]$, one with a test sine and one with a test cosine. In this manner, for each frequency, if the sinusoid of unknown phase is present, but 90 degrees out of phase with one test signal, it is perfectly in phase with the other test signal, thus avoiding the ambiguous correlation value of zero. $A[n]$, then, must be separately correlated with $\sin[2\pi n/N]$ and $\cos[2\pi n/N]$ to adequately detect the presence of $\sin[2\pi n/N + \theta]$, and $A[n]$ must also be separately correlated with $\sin[4\pi n/N]$ and $\cos[4\pi n/N]$ to detect the presence of $\sin[4\pi n/N + \theta]$. If all four correlations (CZLs) are equal to zero, then the signal $A[n]$ is presumed to be equal to zero, assuming that it could only assume the values given in the statement of the problem.

- For a given frequency, and any phase of the unknown signal, correlation values obtained by correlating the unknown with both test cosine and test sine waves having the same frequency as the unknown result that can be used to not only determine the phase of the unknown signal, but to actually completely reconstruct it. This technique is described and illustrated immediately below.

4.3.6   SINGLE FREQUENCY CORRELATION AND RECONSTRUCTION

For a real signal containing a single frequency, the original signal can be reconstructed from correlation values by a simple procedure. For example, if the unknown is

$$x[n] = \sin[2\pi kn/N + \theta]$$

then we compute

$$CZL_{COS} = \sum_{n=0}^{N-1} x[n] \cos[2\pi kn/N] \tag{4.7}$$

and

$$CZL_{SIN} = \sum_{n=0}^{N-1} x[n] \sin[2\pi kn/N] \tag{4.8}$$

where k is a real integer and N is the sequence length; then the original sinusoid can be reconstructed as

$$(1/N)(CZL_{COS} \cos[2\pi kn/N] + CZL_{SIN} \sin[2\pi kn/N]) \tag{4.9}$$

if $k = 0$ or $N/2$, or

$$(2/N)(CZL_{COS} \cos(2\pi kn/N) + CZL_{SIN} \sin(2\pi kn/N)) \tag{4.10}$$

otherwise.

The script (see exercises below)

$$LVxTestReconSineVariablePhase(k1, N, PhaseDeg)$$

allows you to enter a frequency $k1$ and length N to be used to construct three sinusoids, which are 1) a sine wave of arbitrary phase $PhaseDeg$ and frequency $k1$, and 2) a test correlator sine of frequency $k1$, and 3) a test correlator cosine of frequency $k1$.

The call

LVxTestReconSineVariablePhase(1,32,-145)

was used to generate Fig. 4.6, which shows the test correlators in plots (a) and (c), respectively, the sine of arbitrary phase in plot (b), and the perfectly reconstructed sine of arbitrary phase in plot (d).

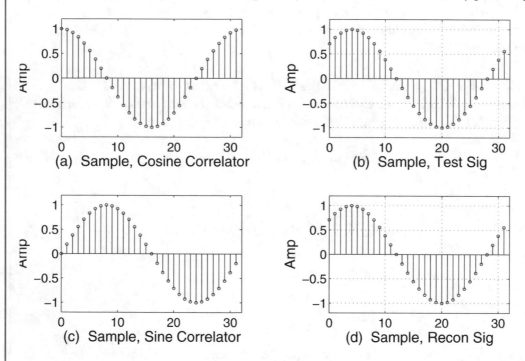

Figure 4.6: (a) and (c) Test Cosine and Sine correlators, respectively; (b) Test sinusoid of arbitrary phase; (d) Test sinusoid reconstructed from the test waveforms at (a) and (c) and the correlation coefficients.

Example 4.8.   Note the fact that the reconstruction formula's weight is 1.0 if k is 0 or $N/2$, but 2.0 otherwise. This is in accordance with the correlation values shown in (4.5) and (4.6). Devise a Command Line call which can be used to illustrate the need for the two different coefficients used in reconstruction.

The following call

$$k = 0; C=\text{sum}(\cos(2*\text{pi}*k*(0:1:7)/8).\hat{}2)$$

can be used, letting k vary as 0:1:7 in successive calls, and noting the value of C for each value of k. The function *cos* in the call should then be changed to *sin*, and the experiments performed again. You will observe that for the cosine function (*cos*), C is N for the case of $k = 0$ or 4 (i.e., $N/2$), but

C is equal to $N/2$ for other values of k. In the case of the sine (*sin*) function, C is zero when k is 0 or $N/2$ since the sine function is identically zero in those cases, and C is $N/2$ for other values of k. This information will be prove to be relevant in the very next section of this chapter as well as in Volume II of the series, where we study the complex DFT.

4.3.7   MULTIPLE FREQUENCY CORRELATION AND RECONSTRUCTION

For a test signal containing a single frequency, a pair of CZLs using cosine and sine waves of the same frequency enables reconstruction of the original test signal using the correlation coefficients and the cosine and sine test waves as basis functions.

This can be extended to reconstruct a complete sequence of length N containing any frequency (integer- or noninteger- valued), between 0 and $N/2$. To achieve this, it is necessary, in general, to do cosine-sine CZLs for all nonaliased integral frequencies possible within the sequence length. For a sequence of length 8, for example, we would do CZLs using cosine-sine correlators of frequencies 0, 1, 2, 3, and 4.

The analysis formulas are

$$R[k] - CZL_{COS}[k] = \sum_{n=0}^{N-1} x[n] \cos[2\pi nk/N] \tag{4.11}$$

$$I[k] = CZL_{SIN}[k] = \mp \sum x[n] \sin[2\pi nk/N] \tag{4.12}$$

where N is the length of the correlating functions as well as the signal $x[n]$, n is the sample index, which runs from 0 to $N - 1$, and k represents the frequency index, which assumes values of $0, 1, .. N/2$ for N even or $0, 1, ...(N-1)/2$ if N is odd.

Equations (4.11) and (4.12) are usually called the Real DFT analysis formulas, and the variables $R[k]$ and $I[k]$ (or some recognizable variation thereof) are usually described as the real and imaginary parts of the (real) DFT.

The synthesis formula is

$$x[n] = (A_k/N) \sum_{k=0}^{K} R[k] \cos[2\pi nk/N] \mp I[k] \sin[2\pi nk/N] \tag{4.13}$$

where $n = 0, 1, ...N - 1$ and $K = N/2$ for N even, and $(N - 1)/2$ for N odd, and the constant A_k =1 if $k = 0$ or $N/2$, and A_k =2 otherwise.

The sign of $I[k]$ in the synthesis formula must match the sign employed in the analysis formula. It is standard in electrical engineering to use a negative sign for the sine-correlated (or imaginary) component in the analysis stage.

Note that the reconstructed wave is built one sample at a time by summing the contributions from each frequency for the one sample being computed. It is also possible, instead of computing one sample of output at a time, to compute and accumulate the contribution to the output of each harmonic for all N samples.

A script (see exercises below) that utilizes the formulas and principles discussed above to analyze and reconstruct (using both methods discussed immediately above) a sequence of length N is

$$LVxFreqTest(TestSeqType, N, UserTestSig, dispFreq)$$

where the argument *TestSeqType* represents a test sequence consisting of cosine and sine waves having particular frequencies, phases, and amplitudes (for a list of the possible parameter values, see the exercises below), N is the desired sequence length, *UserTestSeq* is a user-entered test sequence which will be utilized in the script when *TestSeqType* is passed as 7, and *dispFreq* is a particular test correlator frequency used to create two plots showing, respectively, the test signal and the test cosine of frequency *dispFreq*, and the test signal and the test sine of frequency *dispFreq*. Figure 4.7 shows, for the call

LVxFreqTest(5,32,[],1)

the plots of test signal versus the test correlator cosine having $dispFreq = 1$ at plot (a), and the test correlator sine having $dispFreq = 1$ at plot (b).

Figure 4.8 plots the cosine-correlated and sine-correlated coefficients after all correlations have been performed.

Finally, Fig. 4.9 shows the reconstruction process, in which the coefficients are used to reconstruct or synthesize the original signal. The script actually performs the reconstruction two ways and plots both results together on the same axis: in the first way, one sample at a time of $x[n]$ is computed using formula (4.13), and in the second method, all samples of $x[n]$ are computed at once for each basis cosine and sine, and all weighted basis cosines and sines are summed to get the result, which is identical to that obtained using the sample-by-sample method. The latter method, synthesis harmonic-by-harmonic, gives a more intuitive view of the reconstruction process, and thus the upper plot of Fig. 4.9 shows a 2-cycle cosine, and a 5-cycle cosine, each scaled by the amplitude of the corresponding correlation coefficient and the middle plot shows the same for the sine component, in this case a 1-cycle sine. The lower plot shows the original signal samples as circles, and the reconstructed samples are plotted as stars (since the reconstruction is essentially perfect using both methods, the stars are plotted at the centers of the corresponding circles).

As mentioned above, analysis formulas (4.11) and (4.12) and synthesis formula (4.13) form a version of the Discrete Fourier Transform (DFT) known as the **Real DFT** since only real arithmetic is used. The standard version of the DFT, which uses complex arithmetic, and which has far more utility than the Real DFT, is discussed extensively in Volume II of the series (see Chapter 1 of this volume for a description of the contents of Volume II). In Volume II, the theoretical basis for both the Real and Complex DFTs will be taken up; our brief foray into the Real DFT has served to illustrate the basic underlying principle of standard frequency transforms such as the DFT, which is correlation between the signal and orthogonal correlator pairs of various frequencies.

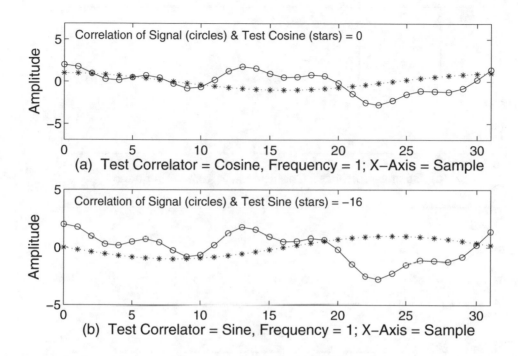

Figure 4.7: (a) Samples of test signal (circles, connected by solid line for visualization of waveform), test cosine correlator (stars, connected with dashed line for visualization of waveform); (b) Samples of test signal (circles, connected with solid line, as in (a)), test sine correlator (stars, connected with dashed line, as in (a)).

4.4   USING ORTHOGONALITY IN SIGNAL TRANSMISSION

The property of orthogonality of sinusoids, i.e., that

$$\sum_{n=0}^{N-1} \cos(2\pi n F/N) \sin(2\pi n F/N) = 0$$

can be put to remarkable use in encoding and transmitting signals. For example, suppose it is desired to transmit two real numbers A and B simultaneously within the same bandwidth. The numbers can be encoded as the amplitudes of *sine* and *cosine* functions having the same frequency F, where F cannot be equal to 0 or $N/2$:

$$S = A\cos(2\pi n F/N) - B\sin(2\pi n F/N)$$

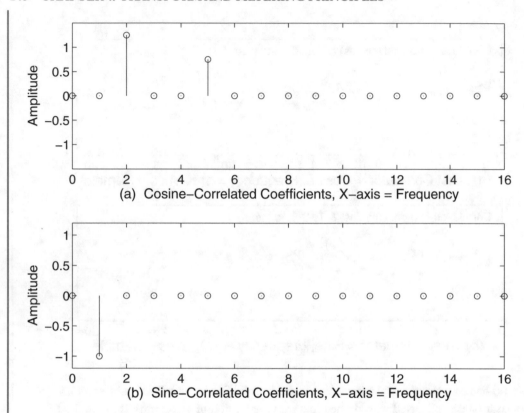

Figure 4.8: (a) Coefficients of cosine-correlated components (scaled according to the synthesis formula); (b) Coefficients of sine-correlated components (scaled according to the synthesis formula).

At the receiving end, because the two carrier waves $\cos(2\pi nF/N)$ and $\sin(2\pi nF/N)$ are orthogonal, it is possible to separately recover the amplitudes of each by multiplying S separately by each, and summing the products (i.e., computing the CZL).

$$S_{R1} = \sum_{n=0}^{N-1} S \cdot \cos(2\pi nF/N)$$

$$S_{R2} = \sum_{n=0}^{N-1} S \cdot \sin(2\pi nF/N)$$

Example 4.9. Let N = 20, F = 7, A = 2 and B = 5. Form the signal S and decode it in accordance with the equations above.

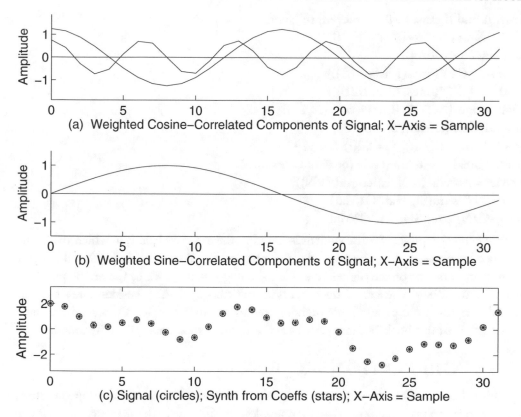

Figure 4.9: (a) Cosine-correlated components, each weighted with its respective coefficient (plotted as continuous functions for ease of visualization); (b) Sine-correlated component, weighted with its coefficient (plotted as continuous function for ease of visualization); (c) Sum of all components, plotted as stars, atop unknown signal's samples, plotted as circles.

A basic script to demonstrate this is

```
function LVOrthogSigXmissBasic(N,F,A,B)
% LVOrthogSigXmissBasic(20,7,2,5)
n = 0:1:N-1; C1 = cos(2*pi*n*F/N);
C2 = sin(2*pi*n*F/N); S = A*C1 - B*C2;
Sr1 = (2/N)*sum(S.*C1)
Sr2 = -(2/N)*sum(S.*C2)
```

A slightly longer script that allows *A* and *B* to be equal length row vectors of real numbers is

```
function [Sr1,Sr2] = LVOrthogSigXmiss(N,F,A,B)
% [Sr1,Sr2] = LVOrthogSigXmiss(20,7,[2,-1,3,0,7],[5,2,-6,3,1])
if ~(length(A)==length(B))
```

```
error('A and B must be the same length'); end
n = 0:1:N-1; C1 = cos(2*pi*n*F/N);
C2 = sin(2*pi*n*F/N);
C1Mat = (C1')*(ones(1,length(A)));
C2Mat = (C2')*(ones(1,length(B)));
AMat = ones(N,1)*A; BMat = ones(N,1)*B;
S1 = C1Mat.*(AMat); S1 = S1(:);
S2 = C2Mat.*(BMat); S2 = S2(:); S = S1 - S2;
% must break S into one cycle (or symbol) frames
SigMat = reshape(S,N,fix(length(S)/N));
Sr1 = (2/N)*sum(SigMat.*C1Mat)
Sr2 = -(2/N)*sum(SigMat.*C2Mat)
```

A more extensive project for the student is found in the exercises below, in which two audio signals are encoded and decoded. The principle of orthogonal signal transmission is found in many applications, from modems on the digital data side to analog systems, such as the briefly proposed FMX system on the 1980s. There are many other types of orthogonal waves besides pure sinusoids, such as Walsh functions, chirps, and pseudorandom sequences, for example. A discussion of these is beyond the scope of this book, but many books and other references may be readily found.

4.5   THE CORRELATION SEQUENCE

The Correlation Sequence of two sequences is obtained by laying one sequence atop the other, computing a correlation (as at the zeroth lag, for example), shifting one of the sequences one sample to the left while the other sequence remains in place, computing the correlation value again, shifting again, etc.

We can define the k-th value of the Correlation Sequence C as

$$C[k] = \sum_{n=0}^{N-1} A[n]B[n+k] \tag{4.14}$$

where k is the Lag Index. If sequences $A[n]$ and $B[n]$ are each eight samples in length, for example, C would be computed for values of k between -7 and +7. Note that in the formula above, $B[n+k]$ is defined as 0 when $n+k$ is less than 0 or greater than $N-1$. Note also that $C[0]$ is the CZL.

Example 4.10.   Use Eq. (4.14) to compute the correlation sequence of the following sequences: $[1, -1]$ and $[-2, 1, 3]$.

The valid index values for each sequence run from 0 to the length of the respective sequence, minus 1. We pick the shorter sequence length to set $N = 2$ and then determine the proper range of k as -1 to +2 (the range of k must include all values of k for which the sequences overlap by at least one sample). Then we get

$$C[-1] = A[0]B[-1] + A[1]B[0] = +2$$

$$C[0] = A[0]B[0] + A[1]B[1] = -3$$

$$C[1] = A[0]B[1] + A[1]B[2] = -2$$

$$C[2] = A[0]B[2] + A[1]B[3] = 3$$

Note that $B[-1]$ and $B[3]$ lie outside the valid index range (0 to N -1) and are defined as having the value 0.

It should be noted that the computations above may be viewed as the multiplication and summing of overlapping samples from the two sequences (or waveforms) when one waveform is held in place and the other slides over the first from right to left, one sample at a time.

Example 4.11. Show graphically how the correlation sequence of two rectangular sequences is computed according to Eq. (4.14).

Figure 4.10, plot (a) shows two sampled rectangles poised, just overlapping by one sample, to compute the correlation sequence. The rightmost one will "slide" over the leftmost one, one sample at a time, and the correlation value at each position is computed by multiplying all overlapping samples and adding all products. In this example, the amplitudes of the two sequences have been chosen so that all products are 1, making the arithmetic easy to do. Plot (b) shows the correlation sequence, plotted for Lag -7, the first Lag index at which the two rectangles overlap. The LabVIEW VIs

DemoCorrelationRectangles.vi

DemoCorrelationSines.vi

and

DemoCorrelationSineCosine.vi

implement correlation, respectively, of two rectangles, two sines, and sine and cosine, on a step-by-step basis, one sample at a time.

A script that implements the same demonstrations as given by the above-mentioned VIs is

ML_Correlation

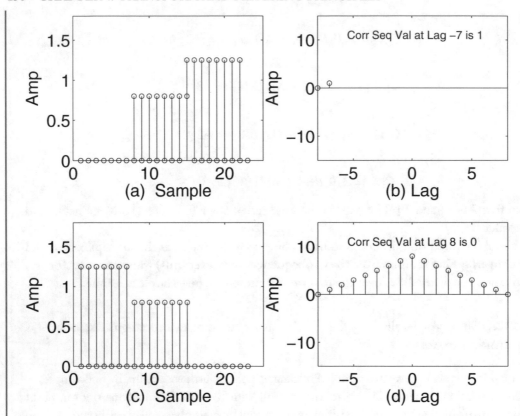

Figure 4.10: (a) Two rectangles just touching, ready to begin computing the correlation sequence; (b) First value of correlation sequence, plotted at Lag -7; (c) The two rectangles, after the one which was right-most originally has been moved to the left one sample at a time to compute the correlation sequence; (d) Complete correlation sequence.

Figure 4.10, plot (c), shows the result after completely "sliding" the rightmost rectangle over the leftmost rectangle, and hence having computed the entire correlation sequence for the two rectangles.

Example 4.12.   Using MathScript, compute and plot the correlation of the sequences $\sin(2\pi(0 : 1 : 7)/4)$ and $\cos(2\pi(0 : 1 : 7)/8)$.

MathScript provides the function *xcorr* to compute the correlation sequence; if two input arguments are provided, the **Cross-correlation** sequence is computed, which is just the correlation sequence as defined in Eq. (4.14). If only one input argument is provided, the correlation of the sequence with itself is computed, which is referred to as the **Auto-correlation** sequence, which is thus defined as

$$C[k] = \sum_{n=0}^{N-1} A[n]A[n+k] \qquad (4.15)$$

We thus run the following m-code:

y = xcorr([sin(2*pi*(0:1:7)/8)],[cos(2*pi*(0:1:7)/8)])
figure; stem(y)

Example 4.13.   Compute and plot the auto-correlation of the sequence $\sin(2\pi(0:1:7)/4)$.

We make the call

$$\textbf{y = xcorr([sin(2*pi*2*(0:1:7)/8)]); figure; stem([-7:1:7],y)}$$

which plots the correlation sequence versus lag number, with 0 representing the two sine sequences laying squarely atop one another without offset, i.e., the CZL. Figure 4.11 shows the results.

Let's compute the correlation sequence between a single cycle of a sine wave having a period of N samples and multiple cycles of a sine wave having a period of N samples. Figure 4.12 shows the correlation sequence generated by correlating a sine wave having a period of 32 samples with four cycles of a sine wave having the same period. Note that the correlation sequence comprises nonperiodic or transient "tail" portions at each end and a central portion that is sinusoidal and periodic over 32 samples.

Now let's consider the correlation sequence between a sine wave having a single cycle over N samples and a sine wave having multiple cycles over N samples. Figure 4.13, shows the result from correlating a single cycle of a sine wave (of period $N = 32$) with a sequence of sine waves having two cycles per N samples. Once the correlation sequence has proceeded to a certain point (the 32nd sample), all samples of the first sequence are overlain with samples of the second sequence. From this time until the second waveform starts to "emerge" (i.e., leave at least one sample of the first sequence "uncovered" or "unmatched"), the correlation sequence value is zero due to orthogonality.

For purposes of discussion, we will refer to the shorter sequence as the correlator, the longer sequence as the test or excitation sequence, and this fully "covered" state as "saturated." In the saturated state, the correlation sequence values reflect a "state-steady" response of the correlator to the test (or excitation) signal when the test signal is periodic.

Figure 4.13 was generated by the script

$$LVCorrSeqSinOrthog(LenSeq1, LenSeq2, Freq1, Freq2, phi)$$

and in specific, the call

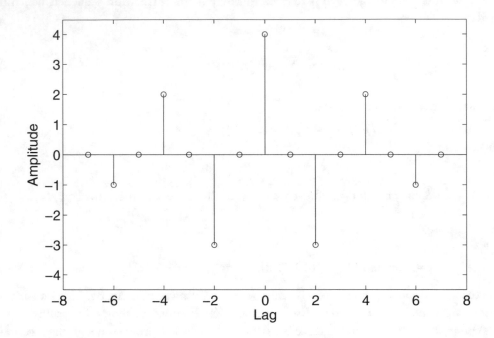

Figure 4.11: The autocorrelation sequence of the sequence sin(2π[0:1:7]/8) plotted against Lag number. Note that the largest value of positive correlation occurs at the zeroth lag when the waveform lies squarely atop itself. This results in every overlying sample pair having a positive product, which in turn results in a large positive sum or correlation value.

LVCorrSeqSinOrthog(32,128,1,8,0)

The script receives as arguments two sequence lengths *LenSeq*1 and *LenSeq*2, two corresponding frequencies *Freq*1 and *Freq*2, and a phase angle for the second sequence. It then constructs two sinusoidal sequences of the lengths, frequencies, and phase specified and computes and plots the correlation sequence.

Let's do the experiment again, with the second sequence having three cycles over the same number of samples as one cycle does in the first sequence. The call

LVCorrSeqSinOrthog(32,128,1,12,0)

yields Fig. 4.14. Again, once the two sequences are in saturation, the output is zero. In this case, the longer sequence (128 samples in all) exhibits a sinusoidal waveform having three cycles over every 32 samples, as opposed to the shorter sequence (32 samples in all), which has only one cycle over its 32 samples. The frequencies (over 32 samples) are thus 1 and 3; they differ by the integer 2.

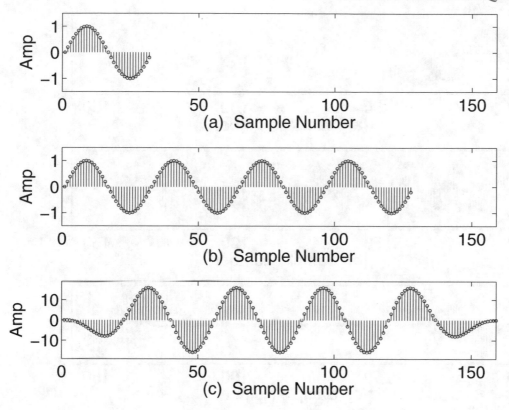

Figure 4.12: (a) First sequence, one period of a sinusoid over 32 samples; (b) Second sequence, four cycles of a sinusoid having a 32-sample period; (c) Correlation sequence, consisting (in the steady state portion) of a sinusoid of period 32 samples.

Example 4.14.   Devise a sequence (i.e., correlator) that will yield steady state correlation sequence values of 0 when correlated with the sequence

[1,0,-1,0,1,0,-1,0,1,0,-1,0]

Observe that the given sequence is in fact the half-band frequency, a sinusoid showing one full cycle every four samples. Sinusoids showing 0 or 2 cycles over four samples will be orthogonal and will yield steady-state correlation values of zero. Two possible sequences are therefore [1,1,1,1] and [1,-1,1,-1]. Two more possible sequences are [-ones(1,4)] and [-1,1,-1,1].

Example 4.15.   Devise several correlators each of which will eliminate from the correlation sequence (in steady state) most of the high frequency information in the following test sequence.

[1,1,1,1,-1,-1,-1,-1,1,-1,1,-1,1,-1,1,-1,1,1,1,1,-1,-1,-1,-1]

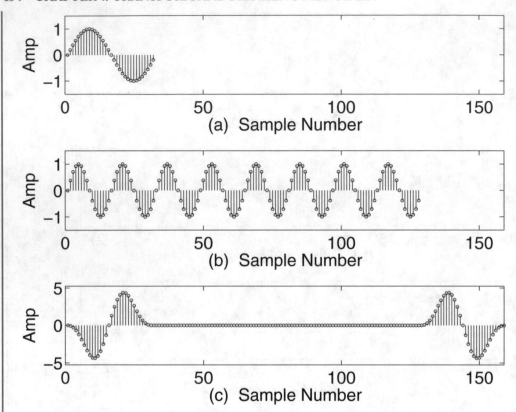

Figure 4.13: (a) First sinusoidal sequence, having one cycle per 32 samples; (b) Second sequence, having two cycles per 32 samples; (c) Correlation sequence of waveforms shown in (a) and (b).

The high frequency information is at the Nyquist limit, and consists of the pattern [1,-1,1,-1]. Correlators such as [1,1] or [-1,-1] (i.e., DC) will eliminate the Nyquist limit frequency. Longer versions also work ([ones(1,4)]) or its negative.

4.6 CORRELATION VIA CONVOLUTION

Previously we've noted that the output of an LTI system can be computed by use of the convolution formula

$$y[k] = \sum_{n=-\infty}^{\infty} x[n]h[k-n]$$

Graphically, this may be likened to time-reversing one of the sequences and passing the left-most sequence (now having negative time indices since it has been time-reversed) to the right

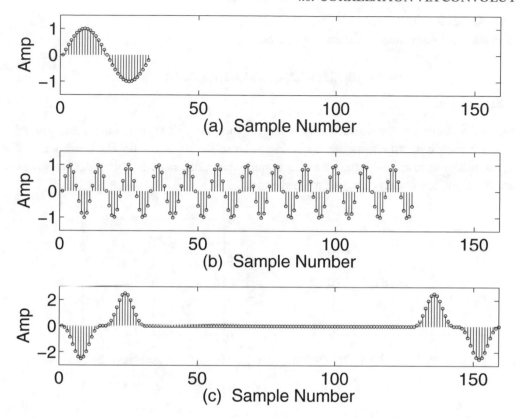

Figure 4.14: (a) First sinusoidal sequence, having one cycle per 32 samples; (b) Second sequence, having twelve cycles per 128 samples, or a net of 3 cycles per 32 samples; (c) Correlation sequence of waveforms shown in (a) and (b).

through the other sequence, sample-by-sample, and computing the sum of products of overlapping samples for each shift.

Correlation, as we have seen above, may be graphically likened to sliding one sequence to the left through the other, summing the products of overlapping samples at each shift to obtain the corresponding correlation value. Correlation can thus be computed as a convolution, by first time-reversing one of the sequences to be correlated prior to computing the convolution.

Example 4.16.   Compute the correlation sequence of [5, 4, 3, 2, 1] and [1, 2, 3, 4, 5] using Math-Script's xcorr function and again using its conv function, and compare the result.

The correlation sequence, obtained by the call

$$\textbf{ycorr = xcorr([5,4,3,2,1],[1,2,3,4,5])}$$

is [25, 40, 46, 44, 35, 20, 10, 4, 1].

The result from convolution, obtained by the call

$$\mathbf{yconv = conv([5,4,3,2,1], fliplr([1,2,3,4,5]))}$$

yields the identical result.

Figure 4.15 illustrates the difference between convolution and correlation for the general case of a nonsymmetric impulse response, while Fig. 4.16 shows the same for the condition of a symmetric impulse response. The latter situation is quite common, since FIR impulse responses, with some exceptions, are usually designed to be symmetric.

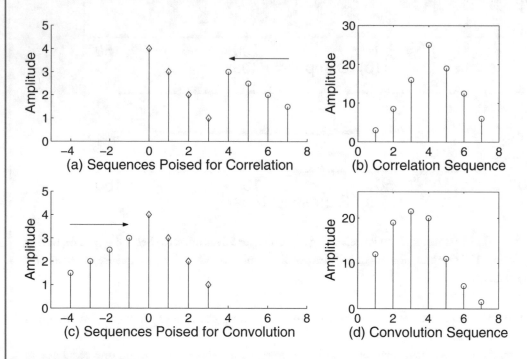

Figure 4.15: (a) Signal sequence, samples 4-7, poised to be correlated with Impulse Response, samples 0-3; (b) Correlation sequence of sequences in (a); (c) Signal, (samples -4 to -1), properly flipped to be convolved with Impulse Response, samples 0-3; (d) Convolution sequence of sequences in (c). The arrows over the signal sequences show the direction samples are shifted for the computation.

Figure 4.17 demonstrates that even though an impulse response may be asymmetric, the convolution and correlation sequences have the same magnitude of chirp response, i.e., the same magnitude of response to different frequencies. Note that there are some differences such as phase and the lead-in and lead-out transients, but the magnitude of the response to different frequencies is essentially the same.

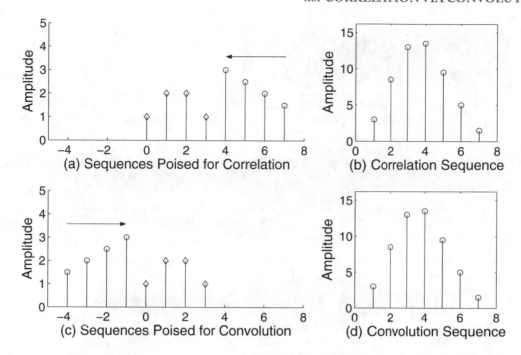

Figure 4.16: (a) Signal sequence, samples 4-7, poised to be correlated with Impulse Response, samples 0-3; (b) Correlation sequence of sequences in (a); (c) Signal, (samples -4 to -1), properly flipped to be convolved with Impulse Response, samples 0-3; (d) Convolution sequence of sequences in (c). Note the symmetrical Impulse Response and hence the identical correlation and convolution sequences.

Note that while the order of convolution makes no difference to the convolution sequence, i.e.,

$$y[k] = \sum_{n=-\infty}^{\infty} b[n]x[k-n] = \sum_{n=-\infty}^{\infty} x[n]b[k-n] \tag{4.16}$$

the order of correlation does make a difference. In general,

$$c_1[k] = \sum_{n=-\infty}^{\infty} b[n]x[n+k]$$

is a time-reversed version of

$$c_2[k] = \sum_{n=-\infty}^{\infty} x[n]b[n+k]$$

In what case would $c_1[k]$ equal $c_2[k]$?

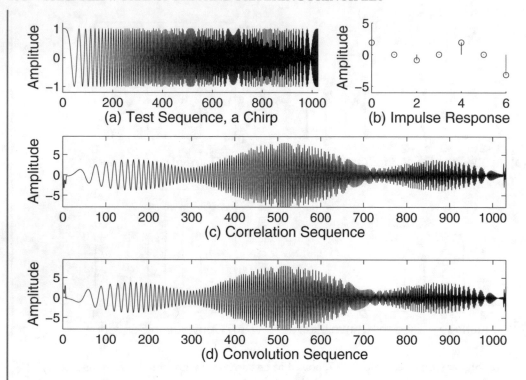

Figure 4.17: (a) Test Linear Chirp, 0 to 512 Hz in 1024 samples; (b) Asymmetric Impulse Response; (c) Correlation Sequence; (d) Convolution Sequence.

4.7   MATCHED FILTERING

Let's consider the problem of detecting an asymmetrically shaped time domain signal, such as a chirp. A good way to detect such a waveform in an incoming signal is to make sure that during convolution, it will correlate well with the impulse response being used. For this to happen, the impulse response will need to be a time-reversed version of the signal being sought. The script

$$LVxMatchedFilter(NoiseAmp, TstSeqLen, FlipImpResp)$$

(see exercises below) illustrates this point. A typical call which reverses the chirp for use as an impulse response in convolution is

LVxMatchedFilter(0.5,128,1)

where the parameter *FlipImpResp* is passed as 1 to reverse the chirp in time, or 0 to use it in non-time-reversed orientation as the filter impulse response. Figure 4.18, plot (a), shows the test sequence, a chirp being received with low frequencies occurring first, containing a large amount of noise, about to be convolved with an impulse response which is the chirp in non-time-reversed format. Plot (b) of the same figure shows the entire convolution sequence, in which no distinctive peak may be seen. Contrast this to the case in which the impulse response is the chirp in time-reversed format, as shown in Fig. 4.19.

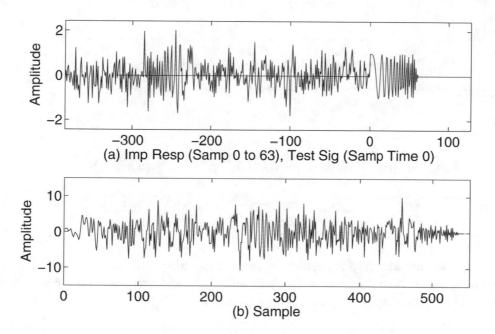

Figure 4.18: (a) Non-time-reversed chirp as impulse response (samples 0 to 63), and incoming signal (samples to left of sample 0); (b) Convolution sequence.

4.8 ESTIMATING FREQUENCY RESPONSE

We've seen above how to determine the frequency content or response of a signal at discrete integral frequencies using Eqs. (4.11) and (4.12). It is also possible to perform similar correlations at as many frequencies as desired between 0 and the Nyquist limit for the signal, leading to a more detailed estimate of the frequency response of a test signal when, for example, considered as a filter impulse response. The following code will estimate the frequency response at a desired number of evenly spaced frequency samples between 0 and the Nyquist limit for the test signal. The loop (four indented lines) may be replaced with the commented-out vectorized code line immediately following. The cosine- and sine-based correlations are done simultaneously with one operation by summing the two correlators after multiplying the sine correlator by $-j$. The result from making the call

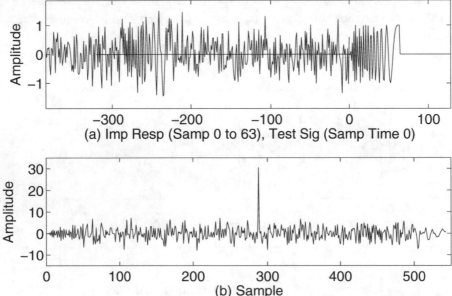

Figure 4.19: (a) Time-reversed chirp as impulse response and incoming signal poised for convolution; (b) Convolution sequence. Note the peak, which occurs when the chirp immersed in noise exactly lays atop the time-reversed chirp impulse response during convolution.

$$\text{LVFreqResp}([1.9,0,-0.9,0,1.9,0,-3.2],500)$$

is shown in Fig. 4.20 (the test signal is the same as that used in Fig. 4.17, in which the frequency response was estimated using a linear chirp).

```
function LVFreqResp(tstSig, NoFreqs)
% LVFreqResp([1.9, 0, -0.9, 0, 1.9, 0, -3.2], 500)
NyqLim = length(tstSig)/2; LTS = length(tstSig);
t = [0:1:(LTS-1)]/(LTS); FR = [];
frVec = 0:NyqLim/(NoFreqs-1):NyqLim;
   for Freq = frVec    % FR via loop
   testCorr = cos(2*pi*t*Freq) - j*sin(2*pi*t*Freq);
   FR = [FR, sum(tstSig.*testCorr)];
   end
% FR = exp(-j*(((2*pi*t)'*frVec)'))*(tstSig');
figure(9)
```

```
xvec = frVec/(frVec(length(frVec)));
plot(xvec,abs(FR));
xlabel('Normalized Frequency')
ylabel('Magnitude')
```

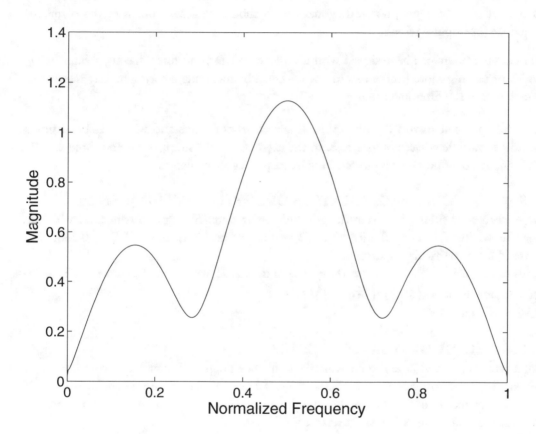

Figure 4.20: The magnitude of frequency response at 500 frequency samples of the sequence [1.9, 0, -0.9, 0, 1.9, 0, -3.2], estimated using cosine-sine correlator pairs at evenly spaced frequencies between 0 and 1.0, normalized frequency.

4.9   FREQUENCY SELECTIVITY

So far, we've seen that:

- A sinusoid of a given period will correlate very well with itself, yielding (in saturation or steady state) a sinusoidal correlation sequence having the same period.

- The correlation sequence of sinusoids that are orthogonal to one another is zero when in steady state.

- A broad principle is that if any particular frequency is desired to be passed from the test sequence to the output sequence with substantial amplitude, that frequency, or a frequency near to it (but, of course, not orthogonal to it!), should be a component of the correlator sequence (or impulse response).

- Very simple filters can be designed with a single correlator, and filters having greater bandwidth or passing more frequencies can be designed by including a corresponding number of correlators in the filter impulse response.

We now proceed to very briefly explore single correlator filters and wider passband filters constructed of multiple adjacent correlators. In the exercises below, simple two- and three-sample lowpass, highpass, bandpass, and bandstop impulse responses are explored.

4.9.1   SINGLE CORRELATOR FILTERS OF ARBITRARY FREQUENCY

Any single-correlator filter of arbitrary length and center frequency can be generated with the following call, where (for example) a length of 54, and normalized frequency of 9/27 = 0.333, have been chosen for this particular experiment.

This results, using a cosine wave as the correlator in the following code, are shown in Fig. 4.21.

```
N = 54; k = 9; x = cos( 2*pi*k*( 0:1:N-1 )/ N );
LVFreqResp(x, 500)
```

4.9.2   MULTIPLE CORRELATOR FILTERS

To make a wider passband covering an arbitrary frequency range, adjacent single correlators can be summed. Adjacent correlators must be 180 degrees out of phase. Note that the amplitude of correlators is adjusted according to whether the frequency is 0, $N/2$, or otherwise, in accordance with our observations made earlier in the chapter.

```
function Imp = LVBasicFiltMultCorr(N,LoK,HiK)
% Imp = LVBasicFiltMultCorr(31,0,7)
Imp = 0;
for k = LoK:1:HiK
if k==0|k==N/2
A = 1; else; A = 2; end
Imp = Imp + A*((-1)^k)*cos( 2*pi*k*( 0:1:N-1 )/ N );
end
LVFreqResp(Imp1, 500)
```

Three filters were generated using the above code with the following calls

```
Imp = LVBasicFiltMultCorr(30,0,4);
```

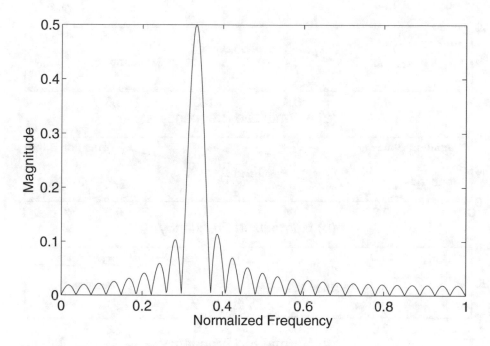

Figure 4.21: The frequency response of a particular single correlator filter.

Imp = LVBasicFiltMultCorr(30,5,9);
Imp = LVBasicFiltMultCorr(30,10,15);

the results of which are shown in Fig. 4.22.

4.9.3   DEFICIENCIES OF SIMPLE FILTERS

Although it is clear that the simple filters we have studied can provide basic filtering responses such as lowpass, etc., there was in actuality very little or no control over a number of parameters that are important. These parameters, which are discussed in detail in Volume III of the series (see the Chapter 1 of this volume for a description of the contents of Volume III), include the amount of ripple (deviation from flatness) in the passband, the maximum response in the stopband(s), the steepness of roll off or transition from passband to stopband, and the phase response of the filter. The lack of adequate signal suppression in the stopbands, is clearly shown in Figs. 4.21 and 4.22, as is the large amount of passband ripple. These and the other deficiencies mentioned will be attacked using a variety of methods to effectively design excellent filters meeting user-given design specifications. The description of these methods requires a number of chapters, but with the brief look we have had at frequency-selective filtering using the fundamental idea of correlation, the reader should have

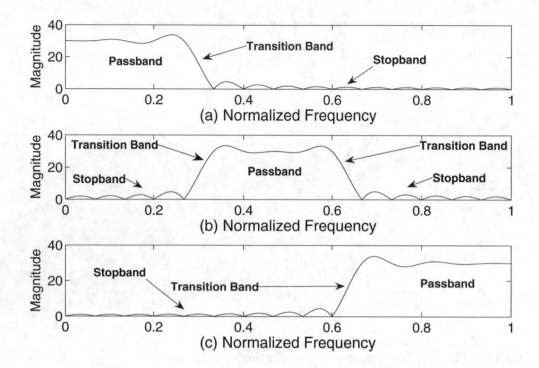

Figure 4.22: (a) Magnitude response of a simple lowpass filter made by summing adjacent cosine correlators; (b) Same, but bandpass; (c) Same, but highpass. For purposes of discussion, a "passband" can be defined as a range of frequencies over which the desired filter response is above a certain level, a "stopband" can be defined as a range of frequencies over which the desired filter response is below a certain level or near zero, and a "transition band" is a range of frequencies lying between a passband and a stopband.

little difficulty understanding the principles of operation and design of FIR filters when they are encountered in Volume III of the series.

4.10   SINUSOIDAL FIDELITY

We noted earlier that in saturation, the correlation sequence between a correlator and a periodic sinusoidal excitation signal contains a steady-state or periodic response in saturation. Now let's consider Fig. 4.23, which shows two sequences, a 32-sample sine wave correlator in plot (a), a 128-sample excitation sequence containing 11.7 cycles of a sine wave in plot (b), and the correlation sequence in plot (c).

Figure 4.23 clearly shows two transient portions, each of which has a length of one sample less than the length of the correlator. The remaining central portion of the correlation sequence

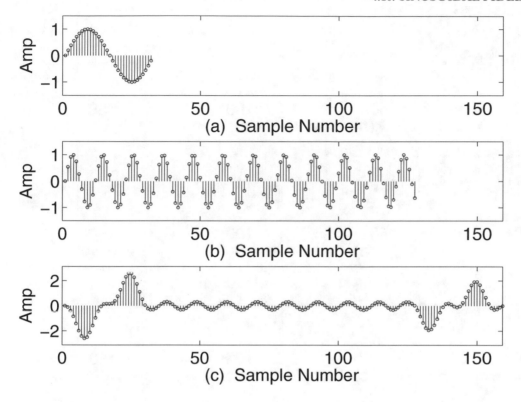

Figure 4.23: (a) Correlator; (b) Test signal; (c) Correlation sequence of sequences shown in (a) and (b).

constitutes the steady-state response. Note that the steady-state response contains only the frequency content of the test signal.

The steady-state portion of the correlation sequence between a sinusoidal excitation signal and a correlator is a sinusoid with the same frequency (or period, in samples) of the sinusoid in the excitation sequence. In other words, what goes in is what comes out, frequency or period-wise, in steady-state. Don't forget that if the two sequences happen to have different integral numbers of cycles over the same period N, the steady-state output will be zero. The correlator need not be sinusoidal, it can be a periodic waveform or random noise, for example, and the same principle holds true, as shown in Fig. 4.24.

This sinusoid-goes-in-sinusoid-comes-out principle is usually referred to as the principle of **Sinusoidal Fidelity**, or **Sinusoidal Invariance**.

By way of contrast, Fig. 4.25 shows the situation when the correlator at plot (a) is a sinusoid and the excitation sequence (at plot (b)) is random noise; the correlation sequence (at plot (c)) is not periodic, since the excitation sequence is not periodic.

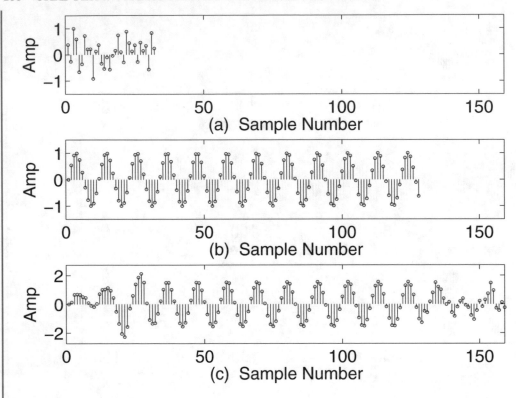

Figure 4.24: (a) Correlator, random noise; (b) Test signal; (c) Correlation sequence of sequences shown in (a) and (b).

The principle of sinusoidal fidelity is only true for sinusoids, which are basic, or fundamental waves which do not break down further into constituent waves. For example, square and sawtooth waves as well as all other periodic waves, are made up of a superposition of harmonically related sinusoids. A sawtooth, for example, requires that its constituent sinusoids have particular amplitudes and that they are all in-phase. In Fig. 4.26, plot (a) shows a correlator, 32 samples of random noise, plot (b) shows a sawtooth, periodic over 12 samples, and plot (c) shows the correlation sequence, in which the sinusoidal constituents of the sawtooth have had their amplitudes and phases (but not their frequencies) randomly shifted by the correlator, resulting in an output waveform having a random shape (peridocity, is, however, retained). For a pure (single-frequency) sinusoidal input, the output waveshape and periodicity are the same as the input, with only the amplitude and phase changing. However, a complex periodic waveform requires that the relative amplitudes and phases of its constituent sinusoids remain the same, otherwise the waveform loses its characteristic shape.

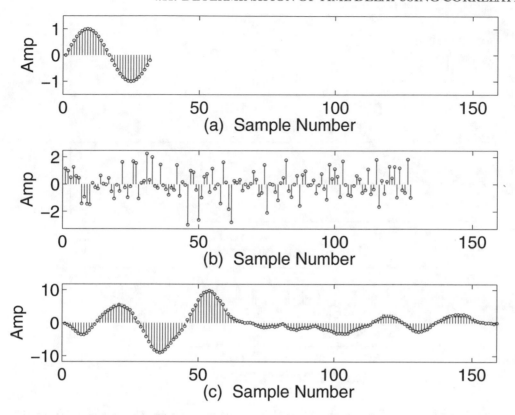

Figure 4.25: (a) Correlator, one cycle of a sine wave over 32 samples; (b) Excitation sequence, 128 samples of random noise; (c) Correlation sequence.

4.11 DETERMINATION OF TIME DELAY USING CORRELATION

Let's take a look at another use for the correlation sequence, estimating the time delay between two sequences.

Figure 4.27 shows the basic setup: two microphones pick up the same sound, but they are at different distances from the sound source. Hence sound features found in the digital sequence from the second microphone will be delayed relative to those same features found in the first microphone's digital sequence. What is sought is the difference in arrival times of the same signal at the two microphones. This information can be used to estimate azimuth (angle) from a point between the microphones to the sound source. This kind of information can be used, for example, to automatically aim a microphone or video camera.

Figure 4.28 depicts two sequences as they were initially captured. The second sequence has several extra delayed versions of the original sound feature (a single cycle of a sine wave) to better

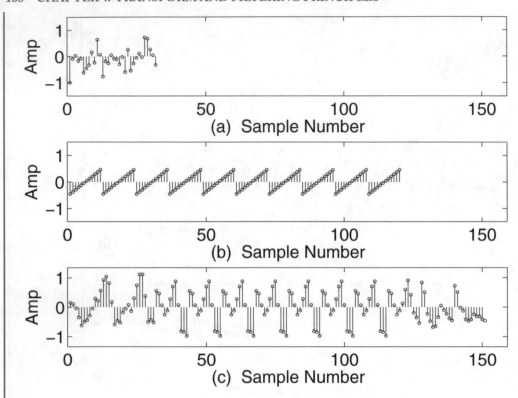

Figure 4.26: (a) Correlator; (b) Test signal; (c) Correlation sequence of sequences shown in (a) and (b).

illustrate the determination of time delay using the cross-correlation sequence. These delayed versions would be analogous to echoes or reflected waves. In a real situation, such echoes would likely be present in the first sequence as well, but they have been left out here so the correlation sequence will be simple to interpret. The use of directional microphones, very common in practice, would also make a significant difference in the content of sequences captured by the two microphones.

The script

LVxCorrDelayMeasure(0.1)

(see exercises below) steps through the computation of the cross-correlation sequence between the two hypothetical captured sound sequences as depicted in Fig. 4.28. The argument in parentheses in the function call specifies the amplitude of white noise added to the basic signal to better simulate a real situation.

Figure 4.29 shows the full cross-correlation sequence, which clearly shows the delays from the principle feature of the first sequence to the corresponding delayed versions found in the second sequence.

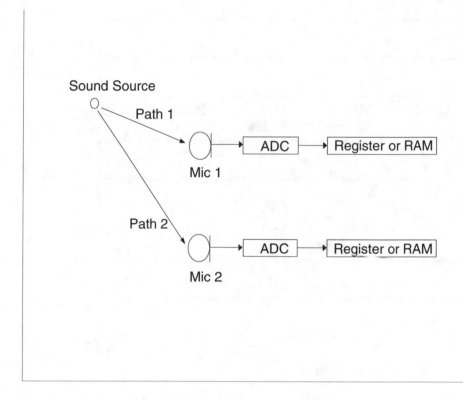

Figure 4.27: Two microphones receiving a sound in a room. The second microphone, further away from the sound source than the first microphone, receives the sounds later. The sounds are digitized and stored in the respective registers.

At Lag 35 (we started out with the two sequences overlapping, i.e., at Lag zero, and started shifting to the left, which increases the Lag index), the main sound feature has reached maximum correlation with the first sequence. At this point, if we know the sample rate, we can compute the time delay and also the path length difference between Paths 1 and 2 from the sound source to the two microphones. For example, if the sample rate were 8 kHz, 35 samples of delay would yield about 0.0044 second of delay. Since sound travels at around 1080 ft/second in air, this implies that the path length difference in feet between paths 1 and 2 was

$$(0.0044 \text{ sec})(1080 \text{ ft/ sec}) = 4.725 \text{ ft}$$

4.12   THE SINGLE-POLE IIR

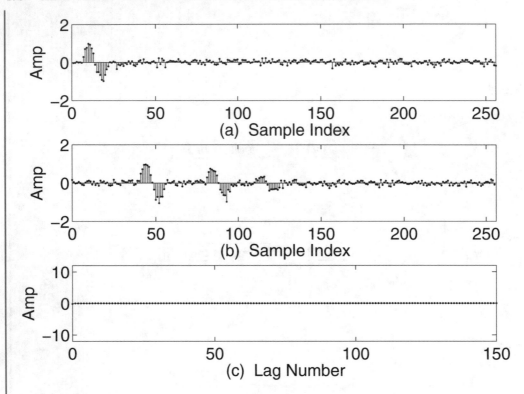

Figure 4.28: (a) First digitized sound; (b) Second digitized sound, not shifted; (c) Correlation sequence (initialized with zeros).

4.12.1 PHYSICAL ARRANGEMENT

Consider the arrangement shown in the left-hand portion of Fig. 4.30. An input signal (shown in plot (a)) enters a summing junction; the sum exits and becomes the output, but the output value also enters a delay, which has an input side and an output side. The value at the input side moves to the output side at every sample or clock time. The large triangle with a number in its interior is a gain block or multiplier, and the number in its interior is multiplied by the delayed output signal, and the product is passed to the summing junction. The value of the gain is called the *pole*, and is in general, a complex number.

4.12.2 RECURSIVE COMPUTATION

In the single pole IIR, the current (or n^{th}) output of the filter is equal to the current (n^{th}) input, weighted by coefficient b, plus the previous (or $(n-1)^{th}$) output weighted by coefficient a (which, for this simple single pole case, is equal to the pole). This can be written as

$$y[n] = bx[n] + ay[n-1] \tag{4.17}$$

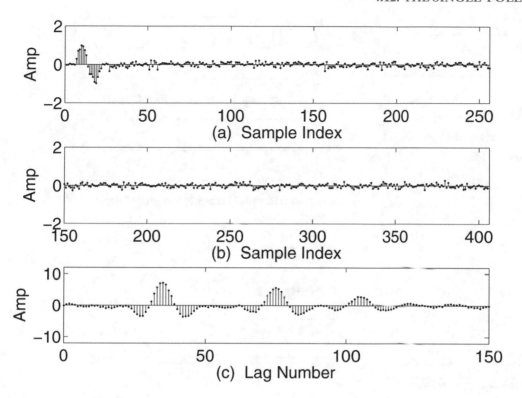

Figure 4.29: (a) First digitized sound; (b) Second digitized sound, shifted 150 samples to the left; (c) Correlation sequence (up to Lag 150).

Example 4.17. Filter the sequence s = $[s_0]$ with a single pole IIR having b = 1 and a = p.

The filter impulse response is $h[n]$ =1, p, p^2, p^3, ...etc. and the output sequence is s_0, ps_0, $p^2 s_0$, $p^3 s_0$... , which is clearly the impulse response weighted by s_0, i.e., $s_0 h[n]$.

If, for example, s_0 = 2, then the output sequence is 2, $2p$, $2p^2$, ...= $2h[n]$.

Example 4.18. Filter the sequence s = $[s_0 \ s_1]$ with the same IIR as used immediately above and show that the output is the superposition of weighted, delayed versions of the filter impulse response.

The first few outputs are

$$s_0 h[0], \ s_1 h[0] + s_0 h[1], \ s_1 h[1] + s_0 h[2], \ ... \qquad (4.18)$$

which can be seen as the sum of two sequences, the second delayed by one sample, namely,

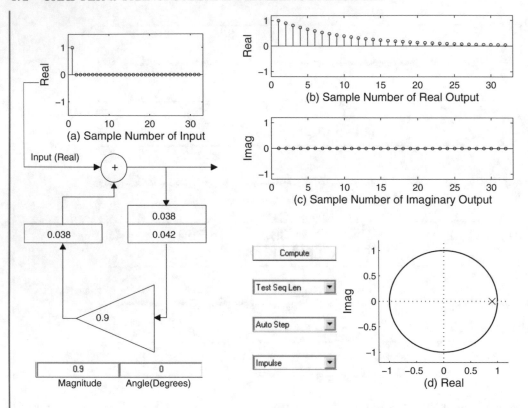

Figure 4.30: (a) Input sequence, a unit impulse; (b) Real part of output sequence, a decaying exponential; (c) Imaginary part of output sequence, identically zero; (d) Complex plane with unit circle and the pole plotted thereon.

$$s_0 h[0], s_0 h[1], s_0 h[2]... = s_0 h[n]$$

and

$$0, s_1 h[0], s_1 h[1], s_1 h[2]... = s_1 h[n-1]$$

and thus it is apparent that the filter output is the superposition of sample-weighted, time-offset versions of the filter impulse response, i.e., the filter output is the convolution of the signal sequence and the filter's impulse response. Output sequence (4.18) can also be visualized as a convolution in which $h[n]$ is flipped from right to left and moved through the signal, i.e.,

$$y[k] = \sum_{n=0}^{1} x[n]h[k-n]$$

where the limits of n have been set appropriately for the signal length.

4.12.3 M-CODE IMPLEMENTATION

The following code shows how to implement the simple difference equation 4.17 in m-code, using values of x, b, p, and SR as shown:

```
SR = 24; b = 1; p = 0.8; y = zeros(1,SR);
x = [1,zeros(1,SR)]; y(1) = b*x(1);
for n = 2:1:SR
y(1,n) = b*x(1,n) + p*y(1,n - 1);
end; figure;
stem(y)
```

The *for* loop, of course, makes this an iterative or recursive computation. Hence the equation serves as a **recursive filter**, a term often used to describe IIR filters.

The function *filter* can also be used to compute the output of a single pole IIR using the following call syntax:

$$\text{Output} = \text{filter}(b,[1,-p],\text{Input})$$

where *Input* is a signal such as (for example) the unit impulse, unit step, a chirp, etc., and b and p are as used in Eq. (4.17).

4.12.4 IMPULSE RESPONSE, UNIT STEP RESPONSE, AND STABILITY

If you are at all familiar with feedback arrangements, you should suspect that if the feedback weight or gain (or in other words, the pole's magnitude) is too large, the filter will become unstable. To remain stable, the magnitude of the pole must be less than 1.0. In the following discussion, we use the impulse and unit step responses corresponding to poles of several magnitudes (< 1.0 and 1.0) to explore the issue of stability in the single-pole IIR.

Impulse Response

Figure 4.30 shows the result from using 0.9 as the value of the pole and an impulse as the input signal. The resultant output, the impulse response, ultimately decays away, and the filter's response to a bounded signal (one having only finite values) is stable. The n-th value of the impulse response $y[n]$ is

$$y[n] = p^n; n = 0 : 1 : \infty$$

- If $|p| < 1$, $|p|^n \to 0$ as $n \to \infty$

- If $|p| = 1$, $|p|^n = 1$ for all n

- If $|p| > 1$, $|p|^n \to \infty$ as $n \to \infty$

Unit Step Response

Figure 4.31 shows the unit step (i.e., DC or frequency 0) response of an IIR with a pole at 0.9. We

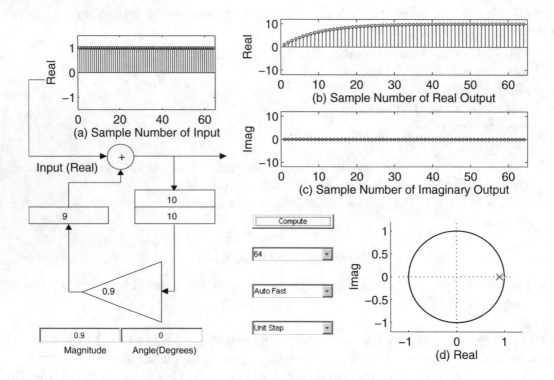

Figure 4.31: (a) Input sequence, a unit step sequence; (b) Real part of output sequence; (c) Imaginary part of output sequence; (d) The pole, plotted in the complex plane.

can determine an expression for the steady-state unit step (DC) response of a single-pole filter by observing the form of the response to the unit step. If the value of the pole is p, then the sequence of output values is $1, 1 + p, 1 + p + p^2$, etc., or at the N-th output

$$y[N] = \sum_{n=0}^{N} p^n \tag{4.19}$$

If $|p| < 1.0$, then $y[N]$ converges to

$$Y_{SS} = \frac{1}{1 - p} \tag{4.20}$$

where Y_{SS} is the response when $N = \infty$, which we'll refer to as the steady state response. In reality, of course, N never reaches infinity, but the difference between the theoretical value of Y_{SS} and $y[N]$

can be made arbitrarily small by increasing N. In the case of Fig. 4.31, the steady state value is

$$Y_{SS} = \frac{1}{1 - 0.9} = 10$$

It can be seen that as p approaches 1.0, the steady state Unit Step response will approach infinity.

Example 4.19.   Demonstrate with several example computations that Eq. (4.20) holds true not only for real poles, but also for complex poles having magnitude less than 1.0.

A simple method is to use the function *filter*; a suitable call for a single-pole IIR is

p = 0.9*j; y = filter([1], [1,-p], [ones(1,150)])

where p is the pole.

Another way is to write an expression which will convolve a unit step of significant length, say 200 samples, with a truncated version of the single-pole IIR's impulse response. A suitable call to create the impulse response and perform the convolution might be

p = 0.9*j; xp = 0:1:99; x = p.^xp; y = conv(x, ones(1,200))

For either of the two methods described above, examine the output sequence to find the steady-state value. If the magnitude of the pole is too close to 1.0, it may be necessary to use longer test sequences than those in the examples above. Once the steady-state value has been found, check it against the formula's prediction. In this case, (for $p = 0.9$*j), both methods produce, after a certain number of samples of output, the steady-state value of $0.5525 + j0.4973$.

Stability

Figure 4.32 shows a single pole filter with borderline stability. In this case, the magnitude of the pole is exactly 1.0, and the impulse response does not decay away. This is analogous to an oscillator, in which a small initial disturbance creates a continuous output which does not decay away. The seriousness of the situation can be seen by using a unit step as the test signal. Figure 4.33 shows the result: a ramp which theoretically would simply continue to increase to infinity if the filter were allowed to run forever.

An accumulator is a digital register having a feedback arrangement that adds the output of the register to the current input, which is then stored in the register. The input signal is thus accumulated or integrated. The single pole IIR with pole value equal to 1.0 functions as an **Integrator**, which we see is not a stable system.

As another demonstration of the visual effect of instability, let's feed a chirp into a filter having a magnitude 1.0 pole (in this case at $\pi/4$ radians, or 45°). Figure 4.34 shows the result: the filter "rings," that is to say, once the chirp frequency gets near the pole's resonant frequency, the output

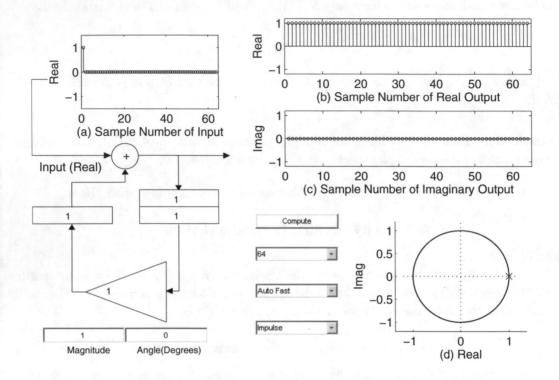

Figure 4.32: (a) Input sequence, a unit impulse sequence; (b) Real part of output sequence, a unit step; (c) Imaginary part of output sequence, identically zero; (d) The pole, plotted in the complex plane.

begins to oscillate and the filter seems to ignore the remainder of the input signal as the frequency passes beyond the pole's resonant frequency.

Reducing the pole's magnitude to 0.95, however, gives the filter a stable response, as shown in Fig. 4.35.

4.12.5 LEAKY INTEGRATOR

A single-pole IIR with a real pole at frequency zero having a magnitude less than 1.0 is often termed a **Leaky Integrator**, and finds frequent use as a signal averager. Often, when the leaky integrator is used to average a signal, the input signal is scaled before it enters the summing junction so that the steady state unit step response is 1.0. The equation of a Leaky Integrator as employed in signal averaging is

$$y[n] = \beta x[n] + (1 - \beta)y[n - 1] \tag{4.21}$$

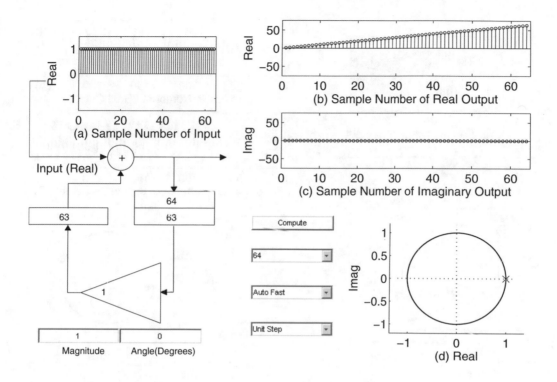

Figure 4.33: (a) Input sequence, a unit step sequence; (b) Real part of output sequence, an increasing ramp; (c) Imaginary part of output sequence, identically zero; (d) The pole, plotted in the complex plane.

where $0 < \beta < 1.0$. Thus in the example above, with the pole at 0.9, we have $(1-\beta) = 0.9$, which implies that $\beta = 0.1$, and the steady state unit step response is 1.0. When used as a signal averager, β is chosen according to the desired relative weights to be given to the current input sample and the past history of input sample values.

Example 4.20.   Verify that Eq. (4.21) will result in a steady state value for y of 1.0 when $\beta = 0.1$ and $x[n]$ is a unit step.

We can use the *filter* function in a straightforward manner a first way

$$x = \textbf{ones(1,100); y = filter(0.1,[1,-0.9],x); figure;stem(y)}$$

or with an equivalent call, which prescales the unit step before filtering

$$x = \textbf{0.1*ones(1,100); y = filter(1,[1,-0.9],x); figure;stem(y)}$$

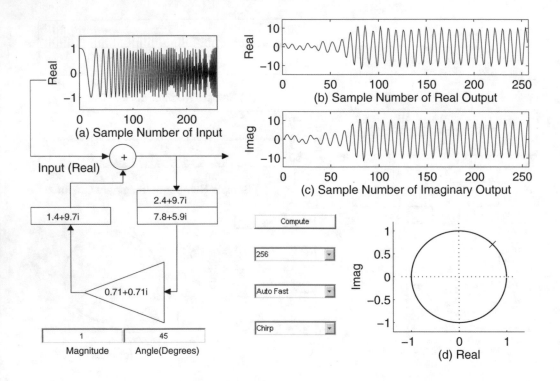

Figure 4.34: (a) Input sequence, a chirp; (b) Real part of output sequence, which is a continuous oscillation at the frequency of the pole; (c) Imaginary part of output sequence; (d) The pole, plotted in the complex plane.

Example 4.21.   Consider the leaky integrator defined by the difference equation $y[n] = 0.1x[n] + 0.9y[n-1]$. For $y[100]$, determine what the relative weights are of $x[100]$, $x[99]$, and $x[98]$ as they appear in an expression for $y[100]$.

By writing out the equations for $y[98]$, $y[99]$, and $y[100]$ and substituting, we get the equation

$$y[100] = 0.1x[100] + 0.09x[99] + 0.081x[98] + 0.72y[97]$$

from which the relative weights can be seen.

4.12.6 FREQUENCY RESPONSE

Whenever the magnitude of the pole is less than 1, the impulse response has a geometrically decaying magnitude. Note that the pole need not be a real number, but a complex number works also, resulting in an impulse response which exhibits a decaying sinusoidal characteristic, with the frequency depending on the pole's angular location along the unit circle—a pole at 0 degrees yields

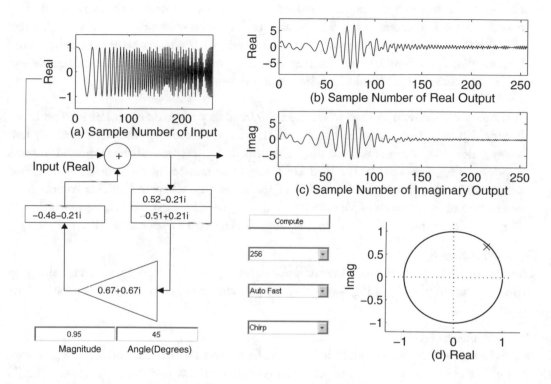

Figure 4.35: (a) Input sequence, a chirp; (b) Real part of output sequence, showing stable resonance at the frequency of the pole; (c) Imaginary part of output sequence; (d) The pole, plotted in the complex plane.

a decaying DC (or unipolar) impulse response, a pole at 180 degrees (π radians) yields the Nyquist frequency (one cycle per two samples), and poles in between yield proportional frequencies. A pole at 90 degrees, for example ($\pi/2$ radians), yields a decaying sinusoidal impulse response exhibiting one cycle every four samples.

- For most filtering projects, the input is real, and it is desired, to keep the hardware and/or software simple, that the output also be real. This implies that the coefficient or pole must also be real. However, for a single feedback delay and multiplier arrangement, the coefficient must be complex if resonant frequencies other than DC and the Nyquist limit frequency are to be had.

- The goal of real input, real coefficients, and a real output, at any frequency from DC to Nyquist can be attained by using one or more pairs of poles that are complex conjugates of each other. This cancels imaginary components in the output signal.

- To select the resonant frequency of a single-pole or complex-conjugate-pair IIR, pick a desired magnitude (< 1.0) and an angle θ between 0 and π radians. Resonance at DC is obtained with $\theta = 0$, resonance at the Nyquist rate is achieved with $\theta = \pi$, resonance at the half-band frequency (half the Nyquist rate) is achieved by using $\theta = \pi/2$, etc. From the magnitude and angle, determine the pole's value, i.e., its real and imaginary parts.

- A simple time-domain method to estimate the frequency response of an IIR is to process a linear chirp. Another method is truncate the IIR's impulse response when its magnitude falls below a designated threshold, and obtain the frequency response of the truncated impulse response using, for example, the Real DFT at a large number of frequencies, as described previously in this chapter. In Volume II of the series (see Chapter 1 of this volume for a description of the contents of Volume II), we'll study use of the z-transform and the Discrete Time Fourier Transform (DTFT) to determine the frequency response of IIRs.

Real Pole at 0 radians ($0°$)
As shown in Fig. 4.36, plot (b), the simple unipolar decaying impulse response, which is a decaying DC signal (as seen in plot (b) of Fig. 4.30) has a lowpass characteristic, although not a particularly good one.

Real Pole at π radians ($180°$)
Changing the pole's value from magnitude 0.9 at 0 radians (0 degrees) to magnitude 0.9 at π radians (180 degrees), yields a net pole value of -0.9. The generates a decaying impulse response $((-0.9)^n)$ which alternates in sign with each sample. This impulse response correlates relatively well with high frequency signals near the Nyquist rate, which also alternate in sign with every sample. Figure 4.37 shows the result; a real pole at -0.9 produces a crude, but recognizable, highpass filter.

Complex Pole
Poles whose angles lie between 0 and π radians generate impulse responses that correlate with various frequencies between DC and the Nyquist limit. Figure 4.38, plot (b), shows the chirp response resulting from selecting the pole at $\pi/2$ radians with a magnitude of 0.9.

- All of Figs. 4.30 through 4.38 were made using the script

$$ML\_SinglePole$$

which creates a GUI with a number of drop-down menus allowing selection of test signal type, manner of computation and display (auto-step, etc.) of output, and test sequence length. Two edit boxes allow entry of the value of the pole using polar coordinates, namely, magnitude and angle in degrees.

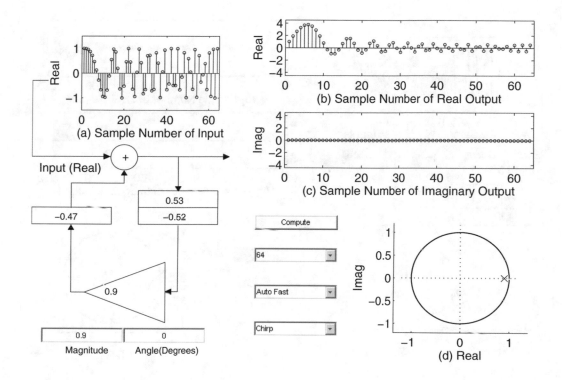

Figure 4.36: (a) Input sequence, a chirp; (b) Real part of output sequence, showing a lowpass effect; (c) Imaginary part of output sequence, identically zero; (d) The pole, plotted in the complex plane.

4.12.7 COMPLEX CONJUGATE POLES

As mentioned above, complex poles can be used in pairs to achieve a filter having real coefficients. Let's start out by determining the net impulse response of two cascaded single pole IIRs having complex conjugate poles. To do this, we convolve the impulse responses of each, which are

$$1, p, p^2, p^3, \ldots$$

and

$$1, p_c, p_c^2, p_c^3, \ldots$$

where p_c is the complex conjugate of p. The first few terms of the convolution are

$$1$$

$$p + p_c$$

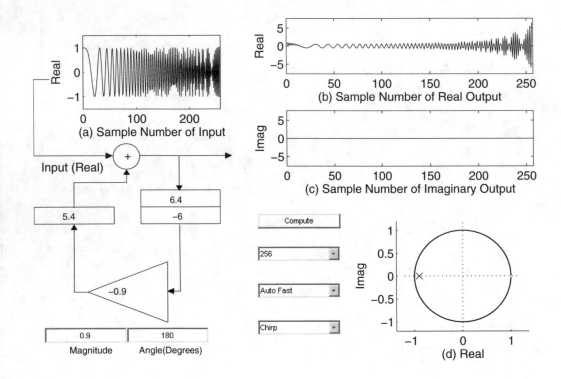

Figure 4.37: (a) Input sequence, a chirp; (b) Real part of output sequence, showing a highpass effect; (c) Imaginary part of output sequence, identically zero.

$$p^2 + pp_c + p_c^2$$

$$p^3 + p^2 p_c + pp_c^2 + p_c^3 \tag{4.22}$$

or in generic terms

$$c[n] = \sum_{m=0}^{n} p^{n-m} p_c^m \tag{4.23}$$

The symmetry of form of the terms of Eq. (4.23) results in a cancellation of imaginary components. For example, we see that

$$p^n + p_c^n = M^n \angle n\theta + M^n \angle(-n\theta)$$

where $p = M\angle\theta$, which reduces to

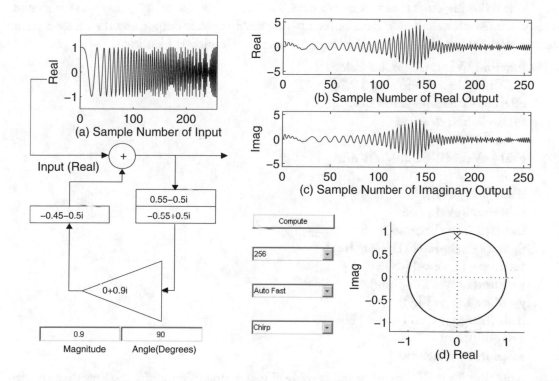

Figure 4.38: (a) Input sequence, a chirp; (b) Real part of output sequence, showing a bandpass effect centered at the halfband frequency; (c) Imaginary part of output sequence, showing a bandpass effect; (d) Unit circle and pole.

$$M^n(\exp(jn\theta) + \exp(-jn\theta)) = 2M^n\cos(n\theta)$$

Similarly, pairs of terms such as

$$[p^2p_c, pp_c^2]$$

sum to real numbers, and

$$pp_c = M\exp(j\theta)M\exp(-j\theta) = M^2\cos(0) = M^2$$

Example 4.22.   Verify that the impulse response of two single pole IIRs having complex conjugate poles is real.

The following script does this in two ways. The first method uses Eq. (4.23), and the second method uses MathScript's *filter* function, computing the first filter's output, and then filtering that result with the second filter having the conjugate pole. The results are identical.

```
function LVImpCmpxConjPoles(P,N)
% LVImpCmpxConjPoles(0.9,24)
cP = conj(P); Imp = zeros(1,N);
for n = 0:1:N; cVal = 0;
for m = 0:1:n
cVal = cVal + (P.^(n-m)).*(cP.^m);
end
if abs(imag(cVal))< 10^(-15)
cVal = real(cVal); end
Imp(1,n+1) = cVal; end
figure(8); subplot(211); stem(Imp)
testImp = [1, zeros(1,N)];
y = filter(1,[1, -P],testImp);
y = filter(1,[1, -cP],y);
if abs(imag(y))< 10^(-15)
y = real(y); end
subplot(212); stem(y)
```

Note that for each case in the script above, due to roundoff error, the imaginary part of the net impulse response still exists, albeit very small.

There is a much better way to compute the output of two cascaded complex conjugate IIRs, and that is by using a single IIR having two stages of delay. The difference equation would be

$$y[n] = bx[n] + a_1 y[n-1] + a_2 y[n-2]$$

We can solve for b, a_1, and a_2 if we know the first few values of the impulse response as computed (for example) by the script above. For $p = 0.9$, for example, the first few values of the impulse response are

$$1, 1.8, 2.43, 2.916, ...$$

Assuming that $y[n]$ is identically zero for $n < 0$, we process a unit impulse ($x[n] = 1$ for $n = 0$, and 0 for all other n) to obtain the impulse response, the values of which we already know from the script above. Thus when $n = 0$, we get $x[0] = 1$, and $y[0] = 1$, requiring that $b = 1$. When $n = 1$, $x[1] = 0$, and $y[1] = 1.8$, which yields $a_1 = 1.8$. For $n = 2$, $x[2] = 0$, $y[2] = 2.43$, and we get $a_2 = -0.81$.

Thus the net difference equation is

$$y[n] = x[n] + 1.8y[n-1] - 0.81y[n-2] \tag{4.24}$$

Note that Eq. (4.24) requires no complex arithmetic at all, and the problem of roundoff error does not occur insofar as imaginary components are concerned. Eq. (4.24) only guarantees, in general, that the first several values of the impulse response will be generated since that is all that has been taken into account. In this case, since the impulse response in question was in fact generated using known poles, and there is no additive noise in the impulse response, the impulse response can be completely generated by solving the difference equation forward in time. In Volume IV of the series (see Chapter 1 of this volume for a description of the contents of Volume IV), we'll investigate this type of process more completely with an algorithm known as Prony's Method, which will allow us to determine a set of coefficients that results in the closest fit to a sequence that might, for example, be quite long and contain noise.

We can use the *filter* function with these coefficients to verify the result

x = [1,zeros(1,50)]; y = filter(1,[1,-1.8,0.81],x);
figure; stem(y)

A yet easier way to obtain the coefficients of the real filter that results from cascading two complex conjugate pole filters is to convolve the coefficient vectors of the two filters, i.e.,

$$Coeff_{CC} = conv([1, -p], [1, -p_c])$$

Example 4.23.   Determine the real coefficients of the second order IIR that results from using the two complex conjugate poles

$$[0.65 + j0.65, 0.65 - j0.65]$$

We convolve the coefficient representations of each:

y = conv([1, -(0.672 +j*0.672)],[1, -(0.672 - j*0.672)])

which yields

y = [1, -1.344, 0.9034]

Since the poles have angles of $\pm\pi/4$ radians, we would expect the peak response of the IIR to lie at one-quarter of the Nyquist rate, or one-eighth of the sample rate. We can explore this with a script that receives one pole as its magnitude and angle in radians, computes the complex conjugate pole, computes the net real coefficients, and filters a linear chirp–a call that yields the answer for the specific problem at hand is

LVRealFiltfromCCPoles(0.95, pi/4)

which results in Fig. 4.39.

```
function LVRealFiltfromCCPoles(PoleMag,PoleAng)
% LVRealFiltfromCCPoles(0.95,pi/4)
Pole = PoleMag*exp(j*PoleAng); cPole = conj(Pole);
rcoeffs = conv([1, -(Pole)],[1, -(cPole)]); SR = 1024;
t = 0:1/(SR-1):1; x = chirp(t,0,1,SR/2);
y = filter([1],[rcoeffs],x); figure(8); plot(y);
xlabel('Sample'); ylabel('Amplitude')
```

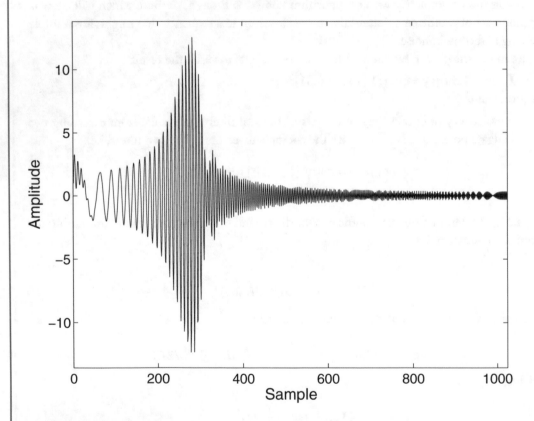

Figure 4.39: Convolution of a linear chirp with a second order all-real coefficient filter obtained by convolving the coefficient vectors of two complex conjugate single pole IIRs.

The results above will be again demonstrated and generalized when the topic of the z-transform is taken up in Volume II of the series. We will also explore different filter topologies or implementations for a given net impulse response.

The LabVIEW VI

DemoDragPolesVI

allows you to select as a test signal a unit impulse, a unit step, or a chirp. The pole or complex conjugate pair of poles is specified by dragging a cursor in the z-plane. From these poles, the VI forms an IIR and filters the selected test signal. The real and imaginary outputs of the filter are plotted. The importance of using poles in complex conjugate pairs can readily be seen by alternately selecting "Single Pole" and "Complex Conjugate Pair" in the Mode Select box. The use of a complex conjugate pair of poles results in all-real filter coefficients and an all-real response to an all-real input signal.

A script that allows you to move the cursor in the complex plane and see the frequency and impulse responses arising from a single pole or a complex-conjugate pair of poles is

ML_DragPoleZero

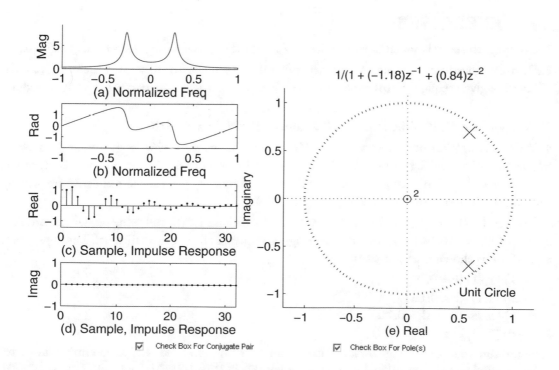

Figure 4.40: (a) Magnitude of frequency response of an LTI system constructed using the poles shown in plot (e); (b) Phase response of same; (c) Real part of impulse response of same; (d) Imaginary part of impulse response of same.

A snapshot of a typical display generated by $ML\_DragPoleZero$ is shown in Fig. 4.40. Note that a pair of complex-conjugate poles is specified and displayed (by the symbol 'X'), and the resulting impulse response, which is real-only, decays since the poles have magnitude less than 1.0.

4.13   REFERENCES

[1] John G. Proakis and Dimitris G. Manolakis, *Digital Signal Processing, Principles, Algorithms, and Applications, Third Edition*, Prentice Hall, Upper Saddle River, New Jersey, 1996.

[2] Alan V. Oppenheim and Ronald W. Schaefer, *Discrete-Time Signal Processing*, Prentice-Hall, Englewood Cliffs, New Jersey, 1989.

[3] James H. McClellan, Ronald W. Schaefer, and Mark A. Yoder, *Signal Processing First*, Pearson Prentice Hall, Upper Saddle River, New Jersey, 2003.

[4] Steven W. Smith, *The Scientist and Engineer's Guide to Digital Signal Processing*, California Technical Publishing, San Diego, 1997.

4.14   EXERCISES

1. Compute the correlation at the zeroth lag (CZL) of the sequences [0.5,-1,0,4] and [1,1.5,0,0.25].

2. Devise a procedure to produce a sequence (other than a sequence of all zeros) which yields a CZL of 0 with a given sequence $x[n]$. Note that there is no one particular answer; many such procedures can be devised.

3. Compute the CZL of the sequences 0.9^n and $(-1)^n 0.9^n$ for n = 0:1:10.

4. Write a simple script to compute the CZL of two sine waves having length N, and respective frequencies of k_1 and k_2; (a) verify the relationships found in (4.6); (b) compute the CZL of two sine waves of length 67 samples, using the following frequency pairs as the respective frequencies (in cycles) of the sine waves: .[1,3],[2,3],[2,5], [1.5,4.5], [1.55,4.55].

5. Use Eqs. (4.7) and (4.8) to derive the correlation coefficients for the following sequences, and then use Eqs. (4.9) and (4.10) to reconstruct the original sequences from the coefficients. The appropriate range for k is shown is parentheses.

        a) [1,-1] (k = 0:1:1)
        b) [1,1] (k = 0:1:1)
        c) [ones(1,4)] (k = 0:1:2)
        d) [1,-1,1,-1] (k = 0:1:2)

6. Determine the correct range for k and use the analysis Eqs. (4.7) and (4.8) to determine the correlation coefficients for the following sequences: [sin(2*pi*(3)*(0:1:7)/8)], [sin(2*pi*(3.1)*(0:1:7)/8)], and [rand(1,8)]. Reconstruct each sequence using the obtained coefficients and Eqs. (4.9) and (4.10).

7. Repeat the previous exercise using signal lengths of 9 instead of 8, and the appropriate range for k.

8. Using paper and pencil, manually compute the auto-correlation sequence of the signal

$$y = u[n] - u[n - 8]$$

9. Use paper and pencil to compute the correlation sequences of the following:

    a) [2,0.5,1,-1.5] and [0.5,2,1,2/3].

    b) [5:-1:1] with itself.

    c) [5:-1:1] and [1:1:5]

10. Manually compute the cross-correlation sequence of the two sequences $\delta[n]$ and $\delta[n - 5]$ over the interval $n = 0{:}1{:}9$. Repeat for the sequences $\delta[n]$ and $u[n - 5]$.

11. Using paper and pencil, use the "sliding waveform method" to compute the convolution and correlation sequences of the following two sequences: [1:1:6] and [2,-1 4,1]. Use the appropriate MathScript functions to verify your results.

12. Verify, using paper and pencil, that the convolution of the sequences [1,2,3,4] and [-1,0.6,3] is the same whichever sequence is chosen to be "flipped" from right to left and moved through the other from the left. Compute the correlation sequence both ways, i.e., picking one sequence, then the other, to be the one that slides from right to left over the other, and verify that the two results are the retrograde of each other.

13. Write a script that will convolve a linear chirp having a lower frequency limit of 0 Hz and an upper frequency limit of 1000 Hz with the following impulse responses and plot the result.

    (a) [0.02,0.23,0.4981,0.23,0.02]

    (b) [0.0087,0,-0.2518,0.5138,-0.2518,0,0.0087]

    (c) [0.5,0,1,0,0.5]

    (d) [0.0284,0,-0.237,0,0.469,0,-0.237,0,0.0284]

    After performing the convolution, state what kind of filtering effect each impulse response has.

14. Repeat the previous exercise, using an upper frequency limit for the chirp of 5000 Hz, and compare the output plots to those of the previous exercise.

15. Repeat exercise 13 or 14 above using a complex chirp rather than a real chirp as the test signal, and plot the magnitude of the convolution as the estimate of frequency response. A complex linear chirp running from frequency 0 Hz to 2500 Hz, for example, may be generated by the following m-code:

```
SR = 5000; t = 0:1/(SR-1):1;
cmpxChrp = chirp(t,0,1,SR/2) + j*chirp(t,0,1,SR/2,'linear',90);
```

16. Plot the convolution of the first 100 samples of the following impulse responses with a linear chirp from 0 to 4000 Hz and characterize the results as to filter type (lowpass, highpass, etc.).

    (a) $h = 0.9^n u[n]5$

    (b) $h = (-0.1^n)0.9^n u[n]$

(c) $h = (1 + (-0.1^n))0.9^n u[n]$

(d) $h = 0.5*exp(j*pi/2).\hat{}(0:8) + 0.5*exp(-j*pi/2).\hat{}(0:8)$

17. Write the m-code for the script

$$LVxMatchedFilter(NoiseAmp, TstSeqLen, FlipImpResp)$$

that generates a test signal of length $TstSeqLen$ consisting of a chirp of a given length (less than $TstSeqLen$) immersed in noise having amplitude $NoiseAmp$ and length $TstSeqLen$ which can be convolved with either the chirp of given length itself or a time-reversed version of it, according to the input argument $FlipImpResp$, which can assume the value 0 to use the non-time-reversed chirp as a filter impulse response or 1 to use the time-reversed chirp as the filter impulse response. Plot the impulse response being used and the test signal, poised to begin convolution, on a single plot, and the convolution sequence on a second plot. The function specification is as follows:

function LVxMatchedFilter(NoiseAmp,TstSeqLen,FlipImpResp)
% Forms a test impulse response, a chirp having a length equal
% to half of TstSeqLen, then builds a test sequence having
% length TstSeqLen and containing the chirp and noise of
% amplitude NoiseAmp. The test signal is then convolved with
% either the chirp or a time-reversed version of the chirp, and
% the results plotted to demonstrate the principle of matched
% filtering.
% FlipImpResp: Use 1 for Time-reversed, 0 for not time-reversed
% impulse response.
% Test calls:
% LVxMatchedFilter(0.5,128,0) % imp resp not time reversed
% LVxMatchedFilter(0.5,128,1) % imp resp time reversed

18. Write the m-code for the script

$$LVxTestReconSineVariablePhase(k1, N, PhaseDeg)$$

receives a frequency $k1$ and length N to be used to construct three sinusoids, which are 1) a sine wave of arbitrary phase $PhaseDeg$ and frequency $k1$, and 2) a test correlator sine of frequency $k1$ and 3) a test correlator cosine of frequency $k1$. Compute the correlation coefficients using Eqs. (4.7) and (4.8), and then reconstruct the sine wave of arbitrary phase $PhaseDeg$ and frequency $k1$ using Eqs. (4.9) and (4.10). Plot the test sine wave, the two correlators, and the reconstructed test sine wave. The function specification is as follows:

function LVxTestReconSineVariablePhase(k1,N,PhaseDegrees)
% Performs correlations at the zeroth lag between
% test cosine and sine waves of frequency k1, having N samples,
% and a sinusoid having a phase of PhaseDegrees, and then

% reconstructs the original sinusoid of phase equal to
% PhaseDegrees by using the CZL values and the test sine
% and cosine.
%Test calls:
% LVxTestReconSineVariablePhase(1,32,45)
% LVxTestReconSineVariablePhase(0,32,90)
% LVxTestReconSineVariablePhase(16,32,90)
% LVxTestReconSineVariablePhase(1,32,45)

19. Write a script that implements the function of the script *LVxFreqTest*. The script should receive the arguments shown in the function definition below, compute the Real DFT coefficients one-by-one using Eqs. (4.11) and (4.12), and then synthesize the original test signal from the coefficients using Eq. (4.13). Three figures should be created, the first one having two subplots and being the analysis window, in which a given set (real and imaginary) of analysis basis functions (i.e., test correlators) having frequency *dispFreq* is plotted against the test signal, and the correlation value according to the Real DFT analysis formula is computed and displayed. The second figure is the synthesis window and should have three subplots, the first two being an accumulation of weighted basis harmonics, and the third being the final reconstruction. The third figure has two subplots and should show the real and imaginary coefficients. Test your script with all six test signals. Examples of each of the three windows are shown in the text.

function LVxFreqTest(TestSignalType,N,UserTestSig,dispFreq)
% TestSignalType may be passed as 1-7, as follows:
% 1 yields TestSig = $\sin(2\pi t) + 1.25\cos(4\pi t + 2\pi(60/360))$ +
% $0.75\cos(12\pi t + 2\pi(330/360))$
% 2 yields TestSig = $\sin(4\pi t + \pi/6)$
% 3 yields TestSig = $\sin(5.42\pi t)$
% 4 yields TestSig = $0.25 + \cos(2.62\pi t + \pi/2) + \sin(5.42\pi t + \pi/6)$
% 5 yields TestSig = $\sin(2\pi t) + 1.25\cos(4\pi t) + 0.75\cos(10\pi t)$
% 6 yields TestSig = $\sin(4\pi t)$;
% 7 uses the test signal supplied as the third argument. Pass this
% argument as [] when TestSignalType is other than 7.
% N is the test signal length and must be at least 2. When
% TestSignalType is passed as 7, the value passed for N is
% overridden by the length of UserTestSig. In this case, N may
% be passed as the empty matrix [] or an arbitrary number if desired.
% dispFreq is a particular correlator frequency which is used to
% create two plots, one showing the test signal chosen by
% TestSignalType and the test correlator cosine of frequency
% dispFreq, and another showing the test signal chosen by
% TestSignalType and the test correlator sine of frequency dispFreq.

```
% Test Calls:
% LVxFreqTest(5,32,[])
% LVxFreqTest(4,19,[])
% LVxFreqTest(7,11,[cos(2*pi*(2.6)*(0:1:10)/11)])
```

20. The script *LV Freq Resp* was presented earlier in the chapter. It evaluated the frequency response of a test signal of length N by correlating the test signal with many test correlators of length N having frequencies evenly spaced between 0 and π radians. In this exercise we develop a method which pads the test signal with zeros to a length equal to a user-desired correlator length, and then performs the Real DFT using the zero-padded test signal

```
function [FR] = LVxFreqRespND(tstSig, LenCorr)
% FR = LVxFreqRespND([ones(1,32)], 128)
% Pads tstSig with zeros to a length equal to LenCorr,
% then computes the Real DFT of the padded tstSig
% over frequencies from 0 to the maximum
% permissible frequency, which is LenCorr/2 if LenCorr
% is even, or (LenCorr-1)/2 if NoFreqs is odd.
% Delivers the output FR as the sum of the real correlation
% coefficients plus j times the imaginary correlation coeffs.
```

For each of the following test signals *tstSig* and corresponding values of *LenCorr*, plot the magnitude of *FR* and compare results to those obtained by performing the same computation using the function *fft*, using the following m-code:

```
figure(55)
subplot(211)
[FR] = LVxFreqRespND(tstSig, LenCorr);
stem([0:1:length(FR)-1],abs(FR))
subplot(212)
fr = fft(tstSig, LenCorr);
if rem(LenCorr,2)==0
plotlim = LenCorr/2 + 1;
else
plotlim = (LenCorr-1)/2 + 1;
end
stem([0:1:plotlim-1],abs(fr(1,1:plotlim)));
Test Signals:
(a) tstSig = ones(1,9); LenCorr = 9
(b) tstSig = ones(1,9); LenCorr = 10
(c) tstSig = ones(1,9); LenCorr = 100
(d) tstSig = ones(1,32); LenCorr = 32
(e) tstSig = ones(1,32); LenCorr = 37
```

(f) tstSig = ones(1,32); LenCorr = 300

21. Write the m-code for the script

$$LVxCorrDelayMeasure(NoiseAmp)$$

Your implementation should have three plots, the first plot being a test signal consisting of a variable amount of noise mixed with a single cycle of a sine wave, which occurs early in the sequence, the second plot consisting of a second test signal consisting of a variable amount of noise and several time-offset cycles of a sine wave, and the third consisting of the cross correlation sequence of the two test signals. The sequence in the second plot should move to the left one sample at a time, and the third plot should be created one sample at a time as the sum of the products of all overlapping samples (i.e., having the same sample index) of the two test signals, as shown on the plots. Figures 4.28 and 4.29 show examples, at the beginning and end of the computation of the correlation sequence, respectively. The function specification is as follows:

function LVxCorrDelayMeasure(k)
% Demonstrates the principle of identifying the time
% delay between two signals which are correlated but
% offset in time. A certain amount of white noise
% (amplitude set by the value of k) is mixed into the signal.
% Typical Test call:
% LVxCorrDelayMeasure(0.1)

22. For an input signal consisting of a unit step, using pencil and paper and the difference equation

$$y[n] = x[n] + ay[n-1]$$

compute the first five output values of a single-pole IIR having the following pole values:

(a) $a = 0.99$
(b) $a = 1.0$
(c) $a = 1.01$

23. Repeat the previous exercise, but, using m-code, compute the first 100 values of output and plot the results on a single figure having three subplots.

24. For an input signal consisting of a unit impulse, using pencil and paper, compute the first five output values of a single-pole IIR having the following pole values:

(a) $a = 0.98$
(b) $a = 1.0$
(c) $a = 1.02$

25. Repeat the previous exercise, but, using m-code, compute the first 100 values of output and plot the results on a single figure having three subplots.

26. Consider a cascade of two single-pole filters each of which has a pole at 0.95. Compute the impulse response of the cascaded combination of IIRs the following two ways:

(a) Compute the first 100 samples of the impulse response of the first IIR, then, using the result as the input to the second filter, compute the net output, which is the net impulse response.

(b) Compute the first 100 samples of the net output of the composite filter made by 1) convolving the coefficient vectors for the two individual single pole filters to obtain a second-order coefficient vector, then 2) use the function *filter* to process a unit impulse of length 100 to obtain the impulse response; compare to the result obtained in (a).

27. Let two IIRs each be a properly-scaled leaky integrator with $\beta = 0.1$. Now construct a filter as the cascade connection of the two leaky integrators. Filter a cosine of frequency 2500 Hz and unity amplitude, sampled at 10,000 Hz with the cascade of two leaky integrators. Determine the steady state amplitude of the output.

28. Determine the correlation (CZL) for the following two sequences, for the values of N indicated, where $n = 0:1:N − 1$:

$$S_1 = \cos(2\pi n3/N) + \sin(2\pi n4/N)$$

$$S_2 = \cos(2\pi n4/N) + \sin(2\pi n3/N)$$

(a) $N = 0.5$;
(b) $N = 1$;
(c) $N = 2$;
(d) $N = 3$;
(e) $N = 6$;
(f) $N = 8$;

29. Write a script that encodes the audio files *drwatsonSR8K.wav* and *whoknowsSR8K.wav* on orthogonal sinusoidal carriers of the same frequency, creates a single signal by taking the difference between the two modulated carriers to create a transmission signal, and then decodes the transmission signal to produce the first audio signal, the second audio signal, or a combination of the two, according to the value of a decoding phase parameter supplied in the function call, according to the specification below. The script should play the decoded audio signal through the computer's sound card. The file *whoknowsSR8K.wav* is much longer than the file *drwatsonSR8K.wav*, so it should be truncated after reading to the same length as *drwatsonSR8K.wav*.

The script should create plots as shown and described in Fig. 4.41:

function LVxOrthogAudio(DecodePhi)
% Creates cosine and sine carriers of equal frequency,
% modulates each by a corresponding audio file,
% drwatsonSR8K.wav or whoknowsSR8K.wav, takes
% the difference to create a transmission signal, then

% decodes the transmission signal to produce one or the
% other of the encoded audio signals, according to the
% value of the variable DecodePhi in the function call.
% DecodePhi is the phase angle of the decoding carrier,
% and should be between 0 and pi/2; 0 will decode the sine
% carrier's audio, pi/2 will decode the cosine carrier's audio,
% and numbers in between 0 and pi/2 will cause a proportional
% mixture of the two audio audio signals to be decoded.
% Test calls:
% LVxOrthogAudio(0)
% LVxOrthogAudio(pi/2)
% LVxOrthogAudio(pi/4)

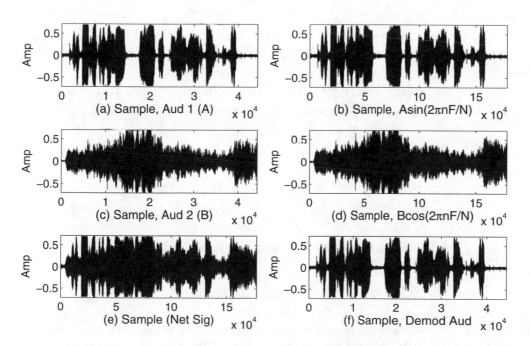

Figure 4.41: (a) First audio signal, *drwatsonSR8K.wav*; (b) Cosine carrier modulated with first audio signal; (c) Second audio signal, *whoknowsSR8K.wav*; (d) Sine carrier modulated with second audio signal; (e) Difference between cosine and sine modulated carriers, forming net transmission signal; (f) Demodulated signal using DemodPhi = 0, yielding the first audio signal.

30. Derive the 2-point impulse responses [1,1] (lowpass) and [1,-1] (highpass) using 2-point cosines. What are the frequencies of the correlators present in each impulse response? Use the script *LVFreqResp*(*tstSig, NoFreqs*) to evaluate the frequency response of each impulse response at 500 points.

31. Derive the 3-point impulse responses [1,0,-1] (bandpass) and [1,0,1] (bandstop) using cosines of length 4. What are the frequencies of the correlators present in each impulse response? Use the script *LVFreqResp*(*tstSig, NoFreqs*) to evaluate the frequency response of each impulse response at 500 points.

32. Develop the script *LVxConvolution2PtLPF*(*Freq*) as described below, and which creates Fig. 4.42, which shows, in subplot (a), the position of the test signal as it moves sample-by-sample from left to right over the two point impulse response, computing the convolution sequence one sample at a

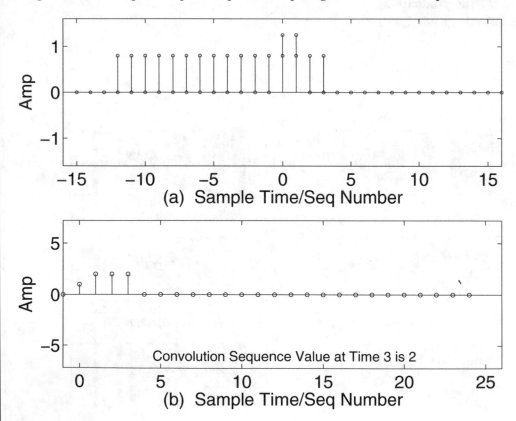

Figure 4.42: (a) Two-sample lowpass impulse response (amplitude 1.25, at sample indices 0 and 1) and DC test signal (amplitude 0.8), advanced from the left to sample index 3; (b) The convolution sequence up to sample index 3.

time, and displaying the convolution sequence in subplot (b). Observe and explain the results from the two sample calls given below.

LVxConvolution2PtLPF(Freq)
% Freq is the frequency of the test sinusoid of length 16 which
% will be convolved with the 2-point impulse response [1, 1].
% Values of Freq up to 8 will be nonaliased.
% Test calls:
% LVxConvolution2PtLPF(0)
% LVxConvolution2PtLPF(8)

33. Develop the script *LVxConvolution2PtHPF(Freq)* as described below, and which creates a figure similar to Fig. 4.42, (except that the impulse response is [1,-1] rather than [1,1]), which shows, in subplot (a), the position of the test signal as it moves sample-by-sample from left to right over the two point impulse response, computing the convolution sequence one sample at a time, and displaying the convolution sequence in subplot (b). Observe and explain the results from the two sample calls given below.

LVxConvolution2PtHPF(Freq)
% Freq is the frequency of the test sinusoid of length 16 which
% will be convolved with the 2-point impulse response [1,-1].
% Values of Freq up to 8 will be nonaliased.
% Test calls:
% LVxConvolution2PtHPF(1)
% LVxConvolution2PtHPF(8)

34. Estimate the frequency response at least 500 frequencies between 0 and π radians of the leaky integrator whose difference equation is

$$y[n] = \beta x[n] + (1 - \beta)y[n - 1]$$

for the following values of β:

(a) 0.01
(b) 0.05
(c) 0.1
(d) 0.5
(e) 0.9
(f) 0.99

35. Compute the convolution sequence of the two sequences [1,1,-1,-1] and [4,3,2,1,4,3,2,1,4,3,2,1,4,3,2,1], identify the transient and steady-state portions of the convolution sequence, and explain the frequency content of the steady-state portion.

APPENDIX A

Software for Use with this Book

A.1   FILE TYPES AND NAMING CONVENTIONS

The text of this book describes many computer programs or scripts to perform computations that illustrate the various signal processing concepts discussed. The computer language used is usually referred to as **m-code** (or as an **m-file** when in file form, using the file extension **.m**) in MATLAB -related literature or discussions, and as **MathScript** in LabVIEW-related discussions (the terms are used interchangeably in this book).

The MATLAB and LabVIEW implementations of m-code (or MathScript) differ slightly (Lab-VIEW's version, for example, at the time of this writing, does not implement Handle Graphics, as does MATLAB).

The book contains mostly scripts that have been tested to run on both MATLAB and Lab-VIEW; these scripts all begin with the letters **LV** and end with the file extension **.m**. Additionally, scripts starting with the letters **LVx** are intended as exercises, in which the student is guided to write the code (the author's solutions, however, are included in the software package and will run when properly called on the Command Line).

Examples are:

LVPlotUnitImpSeq.m

LVxComplexPowerSeries.m

There are also a small number m-files that will run only in MATLAB, as of this writing. They all begin with the letters *ML*. An example is:

ML_SinglePole.m

Additionally, there are a number of LabVIEW Virtual Instruments (VIs) that demonstrate various concepts or properties of signal processing. These have file names that all begin with the letters *Demo* and end with the file extension *.vi*. An example is:

DemoComplexPowerSeriesVI.vi

Finally, there are several sound files that are used with some of the exercises; these are all in the .wav format. An example is:

drwatsonSR4K.wav

A.2 DOWNLOADING THE SOFTWARE

All of the software files needed for use with the book are available for download from the following website:

http://www.morganclaypool.com/page/isen

The entire software package should be stored in a single folder on the user's computer, and the full file name of the folder must be placed on the MATLAB or LabVIEW search path in accordance with the instructions provided by the respective software vendor.

A.3 USING THE SOFTWARE

In MATLAB, once the folder containing the software has been placed on the search path, any script may be run by typing the name (without the file extension, but with any necessary input arguments in parentheses) on the Command Line in the Command Window and pressing *Return*.

In LabVIEW, from the Getting Started window, select MathScript Window from the Tools menu, and the Command Window will be found in the lower left area of the MathScript window. Enter the script name (without the file extension, but with any necessary input arguments in parentheses) in the Command Window and press *Return*. This procedure is essentially the same as that for MATLAB.

Example calls that can be entered on the Command Line and run are

LVAliasing(100,1002)

LV_FFT(8,0)

In the text, many "live" calls (like those just given) are found. All such calls are in boldface as shown in the examples above. When using an electronic version of the book, these can usually be copied and pasted into the Command Line of MATLAB or LabVIEW and run by pressing *Return*. When using a printed copy of the book, it is possible to manually type function calls into the Command Line, but there is also one stored m-file (in the software download package) per chapter that contains clean copies of all the m-code examples from the text of the respective chapter, suitable for copying (these files are described more completely below in the section entitled "Multi-line m-code examples"). There are two general types of m-code examples, single-line function calls and multi-line code examples. Both are discussed immediately below.

A.4 SINGLE-LINE FUNCTION CALLS

The first type of script mentioned above, a named- or defined-function script, is one in which a function is defined; it starts with the word "function" and includes the following, from left to right:

any output arguments, the equal sign, the function name, and, in parentheses immediately following the function name, any input arguments. The function name must always be identical to the file name. An example of a named-function script, is as follows:

function nY = LVMakePeriodicSeq(y,N)
% LVMakePeriodicSeq([1 2 3 4],2)
y = y(:); nY = y*([ones(1,N)]); nY = nY(:)';

For the above function, the output argument is *nY*, the function name is *LVMakePeriodicSeq*, and there are two input arguments, *y* and *N*, that must be supplied with a call to run the function. Functions, in order to be used, must be stored in file form, i.e., as an m-file. The function *LVMakePeriodicSeq* can have only one corresponding file name, which is

LVMakePeriodicSeq.m

In the code above, note that the function definition is on the first line, and an example call that you can paste into the Command Line (after removing or simply not copying the percent sign at the beginning of the line, which marks the line as a comment line) and run by pressing *Return*. Thus you would enter on the Command Line the following, and then press *Return*:

nY = LVMakePeriodicSeq([1,2,3,4],2)

In the above call, note that the output argument has been included; if you do not want the value (or array of values) for the output variable to be displayed in the Command window, place a semicolon after the call:

nY = LVMakePeriodicSeq([1,2,3,4],2);

If you want to see, for example, just the first five values of the output, use the above code to suppress the entire output,and then call for just the number of values that you want to see in the Command window:

nY = LVMakePeriodicSeq([1,2,3,4],2);nY1to5 = nY(1:5)

The result from making the above call is

nY1to5 = [1,2,3,4,1]

A.5   MULTI-LINE M-CODE EXAMPLES

There are also entire multi-line scripts in the text that appear in boldface type; they may or may not include named-functions, but there is always m-code with them in excess of that needed to make a simple function-call. An example might be

N=54; k = 9; x = cos(2*pi*k*(0:1:N-1)/N);
LVFreqResp(x, 500)

Note in the above that there is a named-function (*LVFreqResp*) call, preceded by m-code to define an input argument for the call. Code segments like that above must either be (completely) copied and pasted into the Command Line or manually typed into the Command Line. Copy-and-Paste can often be successfully done directly from a pdf version of the book. This often results in problems (described below), and accordingly, an m-file containing clean copies of most m-code programs from each chapter is supplied with the software package. Most of the calls or multi-line m-code examples from the text that the reader might wish to make are present in m-files such as

McodeVolume1Chapter4.m

McodeVolume2Chapter3.m

and so forth. There is one such file for each chapter of each book, except Chapter 1 of Volume I, which has no m-code examples.

A.6   HOW TO SUCCESSFULLY COPY-AND-PASTE M-CODE

M-code can usually be copied directly from a pdf copy of the book, although a number of minor, easily correctible problems can occur. Two characters, the symbol for raising a number to a power, the circumflex ˆ, and the symbol for vector or matrix transposition, the apostrophe or single quote mark ', are coded for pdf using characters that are non-native to m-code. While these two symbols may look proper in the pdf file, when pasted into the Command line of MATLAB, they will appear in red.

A first way to avoid this copying problem, of course, is simply to use the m-code files described above to copy m-code from. This is probably the most time-efficient method of handling the problem—avoiding it altogether.

A second method to correct the circumflex-and-single-quote problem, if you do want to copy directly from a pdf document, is to simply replace each offending character (circumflex or single quote) by the equivalent one typed from your keyboard. When proper, all such characters will appear in black rather than red in MATLAB. In LabVIEW, the pre-compiler will throw an error at the first such character and cite the line and column number of its location. Simply manually retype/replace each offending character. Since there are usually no more than a few such characters, manually replacing/retyping is quite fast.

Yet a third way (which is usually more time consuming than the two methods described above) to correct the circumflex and apostrophe is to use the function *Reformat*, which is supplied with the software package. To use it, all the copied code from the pdf file is reformatted by hand into one horizontal line, with delimiters (commas or semicolons) inserted (if not already present) where lines have been concatenated. For example, suppose you had copied

```
n = 0:1:4;
y = 2.^n
stem(n,y);
```

where the circumflex is the improper version for use in m-code. We reformat the code into one horizontal line, adding a comma after the second line (a semicolon suppresses computed output on the Command line, while a comma does not), and enclose this string with apostrophes (or single quotes), as shown, where *Reformat* corrects the improper circumflex and *eval* evaluates the string, i.e., runs the code.

$$\text{eval(Reformat('n=0:1:4;y=2.^n;stem(n,y)'))}$$

Occasionally, when copying from the pdf file, essential blank spaces are dropped in the copied result and it is necessary to identify where this has happened and restore the missing space. A common place that this occurs is after a "for" statement. The usual error returned when trying to run the code is that there is an unmatched "end" statement or that there has been an improper use of the reserved word "end". This is caused by the elision of the "for" statement with the ensuing code and is easily corrected by restoring the missing blank space after the "for" statement. Note that the function *Reformat* does not correct for this problem.

A.7   LEARNING TO USE M-CODE

While the intent of this book is to teach the principles of digital signal processing rather than the use of m-code per se, the reader will find that the scripts provided in the text and with the software package will provide many examples of m-code programming starting with simple scripts and functions early in the book to much more involved scripts later in the book, including scripts for use with MATLAB that make extensive use of MATLAB objects such as push buttons, edit boxes, drop-down menus, etc.

Thus the complexity of the m-code examples and exercises progresses throughout the book apace with the complexity of signal processing concepts presented. It is unlikely that the reader or student will find it necessary to separately or explicitly study m-code programming, although it will occasionally be necessary and useful to use the online MATLAB or LabVIEW help files for explanation of the use of, or call syntax of, various built-in functions.

A.8   WHAT YOU NEED WITH MATLAB AND LABVIEW

If you are using a professional edition of MATLAB, you'll need the Signal Processing Toolbox in addition to MATLAB itself. The student version of MATLAB includes the Signal Processing Toolbox.

If you are using either the student or professional edition of LabVIEW, it must be at least Version 8.5 to run the m-files that accompany this book, and to properly run the VIs you'll need the Control Design Toolkit or the newer Control Design and Simulation Module (which is included in the student version of LabVIEW).

APPENDIX B

Vector/Matrix Operations in M-Code

B.1   ROW AND COLUMN VECTORS

Vectors may be either row vectors or column vectors. A typical row vector in m-code might be [3 -1 2 4] or [3,-1,2, 4] (elements in a row can be separated by either commas or spaces), and would appear conventionally as a row:

$$\begin{bmatrix} 3 & -1 & 2 & 4 \end{bmatrix}$$

The same, notated as a column vector, would be [3,-1,2,4]' or [3; -1; 2; 4], where the semicolon sets off different matrix rows:

$$\begin{bmatrix} 3 \\ -1 \\ 2 \\ 4 \end{bmatrix}$$

Notated on paper, a row vector has one row and plural columns, whereas a column vector appears as one column with plural rows.

B.2   VECTOR PRODUCTS

B.2.1   INNER PRODUCT

A row vector and a column vector of the same length as the row vector can be multiplied two different ways, to yield two different results. With the row vector on the left and the column vector on the right,

$$\begin{bmatrix} 1 & 2 & 3 & 4 \end{bmatrix} \begin{bmatrix} 4 \\ 3 \\ 2 \\ 1 \end{bmatrix} = 20$$

corresponding elements of each vector are multiplied, and all products are summed. This is called the **Inner Product**. A typical computation would be

$$[1, 2, 3, 4] * [4; 3; 2; 1] = (1)(4) + (2)(3) + (3)(2) + (4)(1) = 20$$

B.2.2   OUTER PRODUCT

An **Outer Product** results from placing the column vector on the left, and the row vector on the right:

$$\begin{bmatrix} 4 \\ 3 \\ 2 \\ 1 \end{bmatrix} \begin{bmatrix} 1 & 2 & 3 & 4 \end{bmatrix} = \begin{bmatrix} 4 & 8 & 12 & 16 \\ 3 & 6 & 9 & 12 \\ 2 & 4 & 6 & 8 \\ 1 & 2 & 3 & 4 \end{bmatrix}$$

The computation is as follows:

$$[4; 3; 2; 1] * [1, 2, 3, 4] = [4, 3, 2, 1; 8, 6, 4, 2; 12, 9, 6, 3; 16, 12, 8, 4]$$

Note that each column in the output matrix is the column of the input column vector, scaled by a column (which is a single value) in the row vector.

B.2.3   PRODUCT OF CORRESPONDING VALUES

Two vectors (or matrices) of exactly the same dimensions may be multiplied on a value-by-value basis by using the notation " .* " (a period followed by an asterisk). Thus two row vectors or two column vectors can be multiplied in this way, and result in a row vector or column vector having the same length as the original two vectors. For example, for two column vectors, we get

$$[1; 2; 3]. * [4; 5; 6] = [4; 10; 18]$$

and for row vectors, we get

$$[1, 2, 3]. * [4, 5, 6] = [4, 10, 18]$$

B.3   MATRIX MULTIPLIED BY A VECTOR OR MATRIX

An m by n matrix, meaning a matrix having m rows and n columns, can be multiplied from the right by an n by 1 column vector, which results in an m by 1 column vector. For example,

$$[1, 2, 1; 2, 1, 2] * [4; 5; 6] = [20; 25]$$

Or, written in standard matrix form:

$$\begin{bmatrix} 1 & 2 & 1 \\ 2 & 1 & 2 \end{bmatrix} \begin{bmatrix} 4 \\ 5 \\ 6 \end{bmatrix} = \begin{bmatrix} 4 \\ 8 \end{bmatrix} + \begin{bmatrix} 10 \\ 5 \end{bmatrix} + \begin{bmatrix} 6 \\ 12 \end{bmatrix} = \begin{bmatrix} 20 \\ 25 \end{bmatrix} \tag{B.1}$$

An m by n matrix can be multiplied from the right by an n by p matrix, resulting in an m by p matrix. Each column of the n by p matrix operates on the m by n matrix as shown in (B.1), and creates another column in the n by p output matrix.

B.4   MATRIX INVERSE AND PSEUDO-INVERSE

Consider the matrix equation

$$\begin{bmatrix} 1 & 4 \\ 3 & -2 \end{bmatrix} \begin{bmatrix} a \\ b \end{bmatrix} = \begin{bmatrix} -2 \\ 3 \end{bmatrix} \qquad (B.2)$$

which can be symbolically represented as

$$[M][V] = [C]$$

or simply

$$MV = C$$

and which represents the system of two equations

$$a + 4b = -2$$

$$3a - 2b = 3$$

that can be solved, for example, by scaling the upper equation by -3 and adding to the lower equation

$$-3a - 12b = 6$$

$$3a - 2b = 3$$

which yields

$$-14b = 9$$

or

$$b = -9/14$$

and

$$a = 4/7$$

The inverse of a matrix M is defined as M^{-1} such that

$$MM^{-1} = I$$

where I is called the Identity matrix and consists of all zeros except for the left-to-right downsloping diagonal which is all ones. The Identity matrix is so-called since, for example,

$$\begin{bmatrix} 1 & 0 \\ 0 & 1 \end{bmatrix} \begin{bmatrix} a \\ b \end{bmatrix} = \begin{bmatrix} a \\ b \end{bmatrix}$$

The pseudo-inverse M^{-1} of a matrix M is defined such that

$$M^{-1}M = I$$

System B.2 can also be solved by use of the pseudo-inverse

$$\left[M^{-1} \right] [M] [V] = \left[M^{-1} \right] [C]$$

which yields

$$[I] [V] = V = \left[M^{-1} \right] [C]$$

In concrete terms, we get

$$\left[M^{-1} \right] \begin{bmatrix} 1 & 4 \\ 3 & -2 \end{bmatrix} \begin{bmatrix} a \\ b \end{bmatrix} = \left[M^{-1} \right] \begin{bmatrix} -2 \\ 3 \end{bmatrix} \tag{B.3}$$

which reduces to

$$\begin{bmatrix} a \\ b \end{bmatrix} = \left[M^{-1} \right] \begin{bmatrix} -2 \\ 3 \end{bmatrix}$$

We can compute the pseudo-inverse M^{-1} and the final solution using the built-in MathScript function *pinv*:

```
M = [1,4;3,-2];
P = pinv(M)
ans = P*[-2;3]
```

which yields

$$P = \begin{bmatrix} 0.1429 & 0.2857 \\ 0.2143 & -0.0714 \end{bmatrix}$$

and therefore

$$\begin{bmatrix} a \\ b \end{bmatrix} = \begin{bmatrix} 0.1429 & 0.2857 \\ 0.2143 & -0.0714 \end{bmatrix} \begin{bmatrix} -2 \\ 3 \end{bmatrix}$$

which yields $a = 0.5714$ and $b = -0.6429$ which are the same as 4/7 and -9/14, respectively. A unique solution is possible only when M is square and all rows linearly independent.(a linearly independent row cannot be formed or does not consist solely of a linear combination of other rows in the matrix).

APPENDIX C

Complex Numbers

C.1 DEFINITION

A complex number has a real part and an imaginary part, and thus two quantities are contained in a single complex number.

Real numbers are those used in everyday, non-scientific activities. Examples of real numbers are 0, 1.12, -3.37, etc. To graph real numbers, only a single axis extending from negative infinity to positive infinity is needed, and any real number can be located and graphed on that axis. In this sense, real numbers are one-dimensional.

Imaginary numbers are numbers that consist of a real number multiplied by the square root of negative one ($\sqrt{-1}$) which is usually called i or j, and has the property that $i \cdot i = -1$. In this book, j will typically be used to represent the square root of negative one, although either i or j may be used in m-code. Electrical engineers use j as the imaginary operator since the letter i is used to represent current. Typical imaginary numbers might be $5j$, $-2.37i$, etc.

Since complex numbers have two components, they naturally graph in a two-dimensional space, or a plane, and thus two axes at right angles are used to locate and plot a complex number. In the case of the complex plane, the x-axis is called the real axis, and it represents the real amplitude, and the y-axis is called the imaginary axis and numbers are considered to be an amplitude multiplied by the square root of negative one.

Typical complex numbers might be $1 + i$, $2.2 - 0.3j$, and so forth.

C.2 RECTANGULAR V. POLAR

A complex number can be located in the complex plane using either 1) rectangular coordinates (values for horizontal and vertical axes, such as x and y) or 2) polar coordinates, in which a distance from the origin (center of the plot where x and y are both zero) and an angle (measured counterclockwise starting from the positive half of the real or x-axis) are specified.

Figure C.1 shows the complex number $0.5 + 0.6j$ plotted in the complex plane.

You can convert from rectangular coordinates to polar using these formulas:

$$Magnitude = \sqrt{(\text{Re}(W))^2 + (\text{Im}(W))^2}$$

$$Angle = \arctan(\frac{\text{Im}(W)}{\text{Re}(W)}) \tag{C.1}$$

Using $0.5 + 0.6j$ as the complex number, and plugging into the formula for magnitude, we get

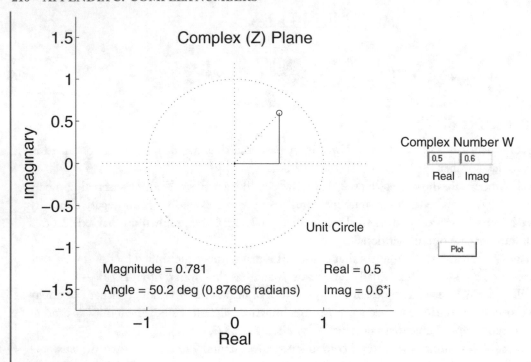

Figure C.1: The complex number -0.5 + 0.6j, plotted in the complex plane.

$$Magnitude = \sqrt{(0.5)^2 + (0.6)^2} = 0.781$$

and the angle would be

$$Angle = \arctan(\frac{0.6}{0.5}) = 0.876\,06 \; radian = 50.2°$$

In m-code, use the function $abs(x)$ to obtain the magnitude or absolute value of x, and use $angle(x)$ to obtain the angle in radians.

Example C.1.   Compute the magnitude and angle of the complex number 0.5 + 0.6*j.

The call

$$x = 0.5 + j*0.6; \; M = abs(x), AngDeg = angle(x)*360/(2*pi)$$

returns M = 0.781 and AngDeg = 50.194.

Example C.2.   Compute the magnitude and angle of the complex number 0.9.

This can be done by inspection since the imaginary part is zero, giving a magnitude of 0.9 and an angle of zero. Graphically, of course, 0.9 is on the positive real axis and thus we also see that its angle must be zero.

Example C.3. Compute the magnitude and angle of the complex number 0 + 0.9j.

By inspection, the magnitude is 0.9 and the angle is 90 degrees or $\pi/2$ radians.

• A common way to express a complex number in polar notation is

$$M \angle \theta$$

where M is the magnitude and θ is the angle, which may be expressed in either degrees or radians.

C.3 ADDITION AND SUBTRACTION

The rule for adding complex numbers is as follows:

$$(a + bj) + (c + dj) = (a + c) + j(b + d)$$

In other words, for addition or subtraction, just add or subtract the real parts and then the imaginary parts, keeping them separate.

C.4 MULTIPLICATION

C.4.1 RECTANGULAR COORDINATES

A real number times a real number is a real number, i.e.,

$$(6)(-2) = -12$$

A real number times an imaginary number is an imaginary number, i.e.,

$$(6)(-2j) = -12j$$

An imaginary number times an imaginary number is negative one times the product of the two remaining real numbers. For example

$$(6j)(-2j) = (6)(j)(-2)(j) = (6)(-2)(j)(j) = (-12)(-1) = 12$$

The general rule is

$$(a + bj)(c + dj) = (ac - bd) + j(ad + bc)$$

C.4.2   POLAR COORDINATES

The product of two complex numbers expressed in polar coordinates is a complex number having a magnitude equal to the product of the two magnitudes, and an angle equal to the sum of the two angles:

$$(M_1 \angle \theta_1)(M_2 \angle \theta_2) = M_1 M_2 \angle (\theta_1 + \theta_2)$$

Example C.4.   Multiply the complex number $0.9 \angle 90$ by itself.

Multiply the magnitudes and add the angles to get $0.81 \angle 180$ which is -0.81.

Example C.5.   Multiply $(0.707 + 0.707j)$ and $(0.707 - 0.707j)$.

$(0.707 + 0.707j) = 1 \angle 45$ and $(0.707 - 0.707j) = 1 \angle -45$ (angles in degrees); hence the product is the product of the magnitudes and the sum of the angles = $1 \angle 0$ = 1. Doing the multiplication in rectangular coordinates gives a product of $(\sqrt{2}/2)(1 + j)(\sqrt{2}/2)(1 - j) = (2/4)(2) = 1$.

C.5   DIVISION AND COMPLEX CONJUGATE

C.5.1   USING RECTANGULAR COORDINATES

The quotient of two complex numbers can be computed by using complex conjugates. The complex conjugate of any complex number $a + bj$ is simply $a - bj$.

Suppose we wished to simplify an expression of the form

$$\frac{a + bj}{c + dj}$$

which is one complex number divided by another. This expression can be simplified by multiplying both numerator and denominator by the complex conjugate of the denominator, which yields for the denominator a single real number which does not affect the ratio of the real and imaginary parts (now isolated in the numerator) to each other.

$$\frac{a + bj}{c + dj} \cdot \frac{c - dj}{c - dj} = \frac{ac - adj + bcj - bdj^2}{c^2 - cdj + cdj - d^2 j^2} = \frac{(ac + bd) + j(bc - ad)}{c^2 + d^2}$$

The ratio of the real part to the imaginary part is

$$\frac{ac + bd}{bc - ad}$$

If a complex number W is multiplied by its conjugate W^*, the product is $|W|^2$, or the magnitude squared of the number W.

$$W W^* = |W|^2$$

As a concrete example, let

$$W = 1 + j$$

Then

$$W^* = 1 - j$$

and the product is

$$1 - j^2 = 2$$

which is the same as the square of the magnitude of W:

$$(\sqrt{1^2 + 1^2})^2 = 2$$

Example C.6.   Simplify the ratio 1/j

We multiply numerator and denominator by the complex conjugate of the denominator, which is $-j$:

$$\left(\frac{1}{j}\right)\left(\frac{-j}{-j}\right) = \frac{-j}{-(-1)} = -j$$

C.5.2 USING POLAR COORDINATES

The quotient of two complex numbers expressed in polar coordinates is a complex number having a magnitude equal to the quotient of the two magnitudes, and an angle equal to the difference of the two angles:

$$(M_1 \angle \theta_1)/(M_2 \angle \theta_2) = (M_1/M_2)\angle(\theta_1 - \theta_2)$$

Example C.7.   Formulate and compute the ratio 1/j in polar coordinates.

The polar coordinate version of this problem is very straightforward since the magnitude and angle of each part of the ratio can be stated by inspection:

$$(1\angle 0)/(1\angle 90) = 1\angle(-90) = -j$$

C.6   POLAR NOTATION USING COSINE AND SINE

Another way of describing a complex number W having a magnitude M and an angle θ is

$$M(\cos(\theta) + j \sin(\theta))$$

This is true since $\text{Re}(W)$ is just $M\cos(\theta)$, and $\cos(\theta)$, by definition in this case, is $\text{Re}(W)/M$; $\text{Im}(W)$ is just $M\sin(\theta)$, and $\sin(\theta) = \text{Im}(W)/M$, so

$$M(\text{Re}(W)/M + j\text{Im}(W)/M) = \text{Re}(W) + j\text{Im}(W)$$

C.7   THE COMPLEX EXPONENTIAL

The following identities are called the Euler identities, and can be demonstrated as true using the Taylor (infinite series) expansions for $e^{j\theta}$, $cos(\theta)$, and $sin(\theta)$:

$$e^{j\theta} = \cos(\theta) + j \sin(\theta)$$

and

$$e^{-j\theta} = \cos(\theta) - j \sin(\theta)$$

where e is the base of the natural logarithm system, 2.718... Such an expression is referred to as a complex exponential since the exponent of e is complex. This form is very popular and has many interesting traits and uses.

For example, by adding the two expressions above, it follows that

$$\cos(\theta) = (e^{j\theta} + e^{-j\theta})/2$$

and by subtracting it follows that

$$\sin(\theta) = (e^{j\theta} - e^{-j\theta})/2j$$

A correlation of an input signal $x[n]$ with both cosine and sine of the same frequency k over the sequence length N can be computed as

$$C = \sum_{n=0}^{N-1} x[n](\cos[2\pi nk/N] + j \sin[2\pi nk/N])$$

or, using the complex exponential notation

$$C = \sum_{n=0}^{N-1} x[n]e^{j2\pi nk/N}$$

The real part of C contains the correlation of $x[n]$ with $\cos[2\pi nk/N]$, and the imaginary part contains the correlation with $\sin[2\pi nk/N]$.

In m-code, the expression

$$\exp(x)$$

means e raised to the x power and hence if x is imaginary, i.e., an amplitude A multiplied by $\sqrt{-1}$, we get

$$\exp(jA) = \cos(A) + j\sin(A)$$

Example C.8. Compute the correlation (as defined above) of $\cos(2\pi(0{:}1{:}3)/4)$ with the complex exponential $\exp(j2\pi(0{:}1{:}3)/4)$.

The call

$$\text{sum}(\cos(2\text{*pi*}(0{:}1{:}3)/4).\text{*exp}(j\text{*}2\text{*pi*}(0{:}1{:}3)/4))$$

produces the answer $2 + j0$.

C.8   USES FOR SIGNAL PROCESSING

- Complex numbers, and in particular, the complex exponential, can be used to both generate and represent sinusoids, both real and complex.

- Complex numbers (and the complex exponential) make it possible to understand and work with sinusoids in ways which are by no means obvious using only real arithmetic. Complex numbers are indispensable for the study of certain topics, such as the complex DFT, the z-Transform, and the Laplace Transform, all of which are discussed in this book.

Biography

Forester W. Isen received the B.S. degree from the U. S. Naval Academy in 1971 (majoring in mathematics with additional studies in physics and engineering), and the M. Eng. (EE) degree from the University of Louisville in 1978, and spent a career dealing with intellectual property matters at a government agency working in, and then supervising, the examination and consideration of both technical and legal matters pertaining to the granting of patent rights in the areas of electronic music, horology, and audio and telephony systems (AM and FM stereo, hearing aids, transducer structures, Active Noise Cancellation, PA Systems, Equalizers, Echo Cancellers, etc.). Since retiring from government service at the end of 2004, he worked during 2005 as a consultant in database development, and then subsequently spent several years writing the four-volume series DSP for MATLAB and LabVIEW, calling on his many years of practical experience to create a book on DSP fundamentals that includes not only traditional mathematics and exercises, but "first principle" views and explanations that promote the reader's understanding of the material from an intuitive and practical point of view, as well as a large number of accompanying scripts (to be run on MATLAB or LabVIEW) designed to bring to life the many signal processing concepts discussed in the series.

Printed in the United States
by Baker & Taylor Publisher Services